CU01163664

Exploiting
Digital Communications

Andrew Richardson

PUBLISHED BY NCC PUBLICATIONS

British Library Cataloguing in Publication Data

Richardson, Andrew
   Exploiting digital communications.
   1. Digital communications
   I. Title
   621.38'0413   TK5103.7

ISBN 0-85102-605-3

© THE NATIONAL COMPUTING CENTRE LIMITED, 1988

All rights reserved. No part of this publication may be reproduced, stored in a retrieval system, or transmitted, in any form or by any means, without the prior permission of The National Computing Centre.

First published in 1988 by:

NCC Publications, The National Computing Centre Limited, Oxford Road, Manchester M1 7ED, England.

Typeset in 10pt Times Roman by H&H Graphics, 2 Duke St, Blackburn; and printed by Hobbs the Printers of Southampton.

ISBN 0-85012-605-3

Acknowledgements

The author wishes to thank numerous manufacturers, users and technical experts for their advice and assistance in the compilation of this book, with particular thanks to:

- Leon Jacob and Mike Birch, Marconi Communications Ltd
- Nick Crowson-Towers, Timeplex Ltd
- Colin Pye, Communications Division, NCC
- Richard Carr, Standards Division, NCC
- Robyn Bray, Consultancy Division, NCC

The Centre acknowledges with thanks the support provided by the Electronics and Avionics Requirements Board of the Department of Trade and Industry for the project from which this publication derives.

Contents

Page

Acknowledgements

How to Use this Book   11

1 Digital Communications   13

2 PTT Carriers and Digital Transmission Services   17

    General Background to the PTTs and Services   17
       The Choice of PTT   17
    Technical Concepts behind Digital Transmission   19
       Digital Transmission Services   19
       PTT Digital Network Infrastructure   22
       North American Digital Services   26
    User Issues Regarding Digital Services   27
       The Benefits of Digital Communications   27

3 Digital Multiplexing   31

    General Background to Digital Multiplexing   31
    Technical Concepts behind Multiplexing   35
       Time Division Multiplexing   35
       Dynamic Time Division Multiplexing   39
       The Pros and Cons of TDM   39
       Statistical Time Division Multiplexing   41
       Features of both TDM and STDM   47

	User Issues Regarding Digital Multiplexers	48
	Integrated Communications	48
	Flexible Control and Resilience	50
	Hardware Redundancy	52
4	**Digital Data Switching**	55
	General Background to Data Switching	55
	Technical Concepts behind Digital Switching	58
	The Megabit Circuit Switch Technology	58
	The Kilobit Matrix Switch Technology	61
	The Data PBX Technology	62
	FEPs and Concentrators	63
	The PABX	65
	User Issues Regarding Digital Switching	67
	ACE in a Corporate Network	67
	The Matrix Solution	68
	The Data PBX Applications	68
	The FEP Solution	70
	The Integrated PABX Solution	71
5	**Digital Communications Applications**	75
	Desktop Video Communications	75
	Megabit SNA Communications	78
	Channel Attachment over Megabit Circuits	79
6	**Packet Switching and the Digital Connection**	83
	General Background to Packet Switching	83
	Technical Concepts behind Packet Switching	84
	The Basic Principles	84
	The X.25 Recommendation	86
	The PAD	87
	The PSE	89
	Packet Switching Gateways	90
	User Issues Regarding Packet Switching	92
	Public Packet Switching Data Networks	92
	Private Packet Switching Data Networks	94
	Local Packet Switching Data Networks	95
	The Switching PAD	96

7	**LANs in a Digital Environment**	99
	General Background to LANs	99
	Technical Concepts behind LANs	100
	LAN Signalling Techniques	100
	User Issues Regarding LANs	102
	LAN to WAN	102
	Digitised Voice Over LANs	104
8	**Digital Network Control and Management**	107
	General Background to the Software Solution	107
	Technical Concepts behind Control and Management	110
	Network Control Features	110
	Network Management Features	115
	The Network Database	117
	User Issues Regarding Network Control and Management	118
	The Network Management System	118
	Other Benefits of a Network Management System	121
9	**The Digital PABX**	123
	General Background to the PABX	123
	Technical Concepts behind Digital PABXs	124
	Digital Signalling Systems	124
	Voice Transmission	126
	The Codec	127
	User Issues Regarding Digital PABXs	128
	The ISPBX	128
	Integrated Services Featurephones	130
	Combining PABXs and LANs	130
10	**The Drive towards ISDN**	133
	General Background to ISDN	133
	In the Beginning . . .	134
	Stage 1: IDN	135
	Stage 2: IDA	136
	Stage 3: ISDN	138
	Stage 4: Broadband ISDN	138
	Technical Concepts behind ISDN	140

	IDA	140
	ISDN	142
	User Issues Regarding ISDN	145
	The IDA Experience	145
	The Dawn of ISDN	150

11 Digital Implementation 153

Installing Point-to-point Digital Circuits	153
Point-to-point Circuits	154
Single Channel Synchronous Data	154
Single Channel Asynchronous Data	156
Single Channel Voice	156
Multi-channel Point-to-point Data	156
Multipoint Circuits	157
Data Circuit Resilience	158
Data and Voice Switching	158
Public and Private ISDN	159
Management Integration	159

Appendix

1 Glossary	161
2 Bibliography and References	169
3 Summary of CCITT G-series Recommendations	171
4 Summary of CCITT I-series Recommendations	173

Index 175

How to Use this Book

Digital Communication is expected to have an explosive impact on the world of telecommunications now and in the foreseeable future. It is the purpose of this publication to cover the advantages and disadvantages of this developing technology; to show how and where it can best be exploited; to provide an overview of the equipment currently available for digital communication and how it can best be applied; and generally to identify important considerations for formulating a successful digital migration strategy.

The book itself is divided into a number of chapters each of which is devoted to a specific topic. Where possible, each chapter is then subdivided into three sections:

General background, a general introduction to the topic suitable for readers from a variety of backgrounds;

Technical concepts which goes into technical detail on the particular topic, and is designed for the more technical user with some grounding in communications technology;

User issues which covers the business advantages and disadvantages of the topic, together with possible applications and uses, and is again suitable for general readers.

Each section has been further subdivided where possible to aid clarity and to allow for easy reference via the contents page.

As with most technical disciplines, a special terminology has evolved in parallel with the evolution of digital communication. This terminology has been complicated by the fact that manufacturers and suppliers tend in many cases to use different terms and phrases to describe the same concept or device. Where several terms exist I have therefore chosen the most appropriate, and attempted to use this term throughout the publication.

Below are a few of the terms used in this publication, with the variations as given by different manufacturers and suppliers. Other terms used later will be explained as they occur, or can be checked in the glossary.

High-speed circuit: This basically covers any communications link capable of transmitting at speeds of thousands or millions of bits per second, these speeds being expressed as *kilobit* or *megabit transmission*. The circuit itself usually forms the physical communications medium between two or more communications devices (eg a high-speed circuit linking two remotely located computers). In the various manufacturers' literature, a high-speed circuit may often be referred to as the composite link, aggregate link, inter-node circuit, data link, bearer circuit or similar.

Low-speed circuit: This will generally refer to any communications link operating at speeds of less than 19.2 kilobits per second. Such a link could be either digital or analogue in nature, and would normally be used in reference to a 'tail-end' communications link between, for example, a multiplexer and a terminal. In such a case a number of low-speed terminal links may well be multiplexed together onto a single high-speed circuit. In manufacturers' literature low-speed circuits may often be referred to as low-speed channels, low-speed interfaces, data ports, etc.

Kbits and Mbits: Both of these abbreviations will follow a numerical figure and will be used to express the bit transmission rate of data over a particular circuit. Kbits indicates 'thousands of bits per second' (often expressed in general literature at 'kbps' or 'kbit/s'), while Mbits indicates 'millions of bits per second' (otherwise expressed as 'Mbps' or 'Mbit/s'). In addition I have expressed 2.048 Mbits speeds as 2 Mbits for clarity.

Node: This term refers to a device within a digital network which acts as an access point to users connected via low-speed circuits, and/or provides switching facilities for circuits, channels or data passing through it. Nodes include devices such as switching multiplexers and data PBXs.

1 Digital Communications

By the beginning of the next century the United Kingdom, along with the rest of Western Europe, North America and most other technologically developed nations, will have made the transition from a limited-function, analogue-based telecommunications network to an integrated digital communications system capable of supporting a wide variety of business and residential communication needs.

Even today there are a growing number of digitally based services and facilities on offer to the commercial subscriber, any one of which could have a profound effect not only on the corporate communications structure but also, ultimately, on an organisation's business strategy. In order for individual organisations within the UK to take full advantage of these developments, it is important that those responsible for corporate communications strategy should begin to consider seriously the future implications, applications and implementations of digital technology within their organisation. In particular, consideration should be given to when, where, how and indeed if such technology is to be introduced. The most important factor in deciding such issues may very well be that of cost. Potential cost savings and benefits must be weighed against what may very well be a large capital investment programme. But what are the benefits of digital communications?

Until recently, telecommunications were based on 19th-century analogue technology which amongst other drawbacks was slow in operation, liable to circuit noise and interference, fault-prone and generally outdated in regard to the limited, disjointed services on offer. Digital communications, on the other hand, offer a way of representing voice, data, text and image in a standard format over existing physical media (such as twisted pair wiring) or over the new media of optic fibre, microwave or satellite links, all of which offer great improvements in speed, bandwidth and quality. Digital communication also integrates with the digital technologies of data processing and office automation, which together encompass the concept of 'Information Technology'. This standardisation and integration has resulted in new services and benefits being provided to the business community.

The impact on business practices can be fairly dramatic. Already a number of organisations are moving away from the traditional 'person-to-person' communication (using face-to-face meetings or the standard telephone handset) towards communication based on integrated combinations of voice, data, text and image transmission between personal multi-purpose desktop workstations. Such an application highlights the future potential of digital communication and its capacity to alter corporate working patterns in the near future.

When considering the potential of digital communications, there are a number of salient facts to take into consideration:

— In most organisations there will be a period of some years during which existing analogue networks and new digital networks will have to co-exist in some form of hybrid structure. Interworking, control and management will be obvious problems in such a case.

— Analogue equipment in such cases will be phased out rather than immediately discarded, mainly due to the substantial capital investment in existing analogue equipment.

— As digital communications technology is still in a comparatively early stage of development, decisions about what, when and where to purchase will be important factors in equipment acquisition.

— Networking standards and services are still evolving, so current digital equipment must be flexible enough to meet future requirements.

— Digital communication equipment and services are still on the whole more costly than their analogue counterparts.

Along with such considerations, digital communications have brought a new set of potential problems above and beyond those encountered in analogue networks. These can be summarised as follows:

DIGITAL COMMUNICATIONS

- The integration and expansion of digital communication services within any organisation means that the network must be controlled and managed as a strategic company resource, in a similar manner to resources such as capital, plant and manpower.

- The digital communications revolution means the introduction of more complex and costly equipment, but provides benefits in the form of a growing number of effective and cost-efficient services and facilities.

- Specialist staff with in-depth knowledge of digital communications are increasingly more scarce than the average communications technician, and hence more costly.

- Some of the network management functions associated with analogue circuits now come under the control of a digital circuit provider, so the subscriber effectively leases a 'managed' digital circuit. This has plus and minus points.

- The cost of leased digital circuits (especially those of megabit operation) makes the standby circuit option relatively expensive, yet a failure in a digital link can disrupt the work of large numbers of users.

All these problems can, however, be solved, and in any case the potential for substantial cost savings and enhanced capabilities may outweigh any of these drawbacks for many organisations. Other organisations, although recognising the benefits, may feel the time is not yet right for such innovative moves into digital communications, but are nevertheless ensuring that decisions taken today do not limit their future choices for implementing digital communications technology.

2 PTT Carriers and Digital Transmission Services

GENERAL BACKGROUND TO THE PTTs AND SERVICES
The Choice of PTT

Within the United Kingdom, there are at the time of writing three government-licensed PTT carriers who supply telecommunications circuits and related services to subscribers. These carriers are British Telecommunications plc, Mercury Communications Ltd and (in its own geographical region) Hull City Council Telephone Department. All three carriers have faced the task of developing a digitally based network, although each encountered a different set of problems and obstacles in achieving its goal.

Because of its size, complexity and heavy capital investment in older technologies, British Telecom faced a mammoth task when coverting from analogue to digital operation. It was obvious that such a task would take many years and millions of pounds to complete. In the interim period it would have to maintain interconnection between the digital and analogue networks, providing enough temporary interfaces between the growing digital and declining analogue network structures, but in a cost-effective manner. Hull City Council had a similar exercise although on a very much smaller scale. For both organisations, forward planning and design were important stages in the overall conversion project.

Mercury had a definite advantage over the established PTTs in that the company could literally 'start from scratch' using the most up-to-date digital technology without the need to consider any existing analogue network commitments. This gave flexibility in the initial digital network design, although the return on investment would not be realised until the network was available over most of the UK. This meant that service availability was of paramount importance.

The following is a brief overview of the major carriers' involvement in digital networking.

British Telecommunications plc

The British Telecommunications Act 1981 set into motion the beginnings of telecoms liberalisation within the UK. The Act brought British Telecom into being. The organisation was formed to take over telecommunications and data processing operations from the Post Office.

In August of that same year, British Telecom brought the first of its digital 'X-Stream' services into operation. Originally marketed as the 'Packet Switching Service' but later renamed 'Packet SwitchStream' (PSS), the service was designed to provide the subscriber with a common network access point in the form of standard CCITT interfaces which could connect to a variety of end-user data equipment. PSS also provided connections out into the international packet switching network through 'International Packet SwitchStream' (IPSS).

In 1982, the 2 Mbits digital service 'MegaStream' became available on a limited basis around the country, and was followed in mid-1983 by the 64 Kbits 'KiloStream' service. Both these services provided the subscriber with a non-switchable, point-to-point high-speed circuit, with the ability to carry either voice traffic, data traffic, or a combination of both when used in combination with digital multiplexers.

All these early digital services were developed to give user organisations some of the benefits and facilities of digital communications, but in many ways could be viewed as interim solutions until the arrival of the Integrated Services Digital Network (ISDN) on a national and eventually international basis.

The first pilot trials of ISDN in the form of Integrated Digital Access (IDA) began in mid-1986. This service was seen as evolving a route to full ISDN facilities, and was therefore complementary to, as well as providing interworking with, most of the existing analogue and digital services.

Mercury Communications Ltd

The 1981 Act also permitted the Secretary of State to licence other organisations to provide telecommunications services. Such a licence was granted to a consortium which in 1982 formed the company Mercury Communications.

This alternative national public carrier service initially installed a geographic 'figure of eight' network consisting of optical fibre cable laid in ducts alongside British Rail tracks. This enabled most of the major commercial centres in England to be linked. Microwave radio, operating from a number of main distribution points, furnished local connection to customers. Additional links to other major cities in the UK were then established by the use of microwave trunk routes. During the development phase Mercury supplied point-to-point leased lines only. This was essentially a technical constraint brought about by the necessity to build a substantial network before the introduction of switched services.

Users of the initial service have a number of alternative ways to link into the network. With the installation of a microwave dish on their premises, many users connect via microwave radio links to the nearest Mercury microwave node. The proviso in using microwave is that the user is within 25 km of the node for a 2 Mbits link, or within 10 km for 64 Kbits link, and that a 'line of sight' is available between the microwave dish and the node. Alternative methods of connection to the network now include intra-city coverage via cellular microwave radio, the use of optical fibre networks, or interconnection via third party networks such as cable TV systems.

Microwave spur links are still being established by Mercury: these will expand the network to give nationwide coverage. Satellite links give the network its international dimension, as well as providing additional UK coverage.

The Mercury network is designed to carry all forms of digital services, including the capability to handle digital switched services (marketed as the 'Mercury 2000' services). This provides the base on which a fully fledged ISDN system can be developed.

TECHNICAL CONCEPTS BEHIND DIGITAL TRANSMISSION
Digital Transmission Services

Under the auspices of the 'Conference of European Posts and Telecommunications administrations' (CEPT) the European PTTs have formulated a digital hierarchy for voice circuit standards within Europe. These standards are:

64 Kbits	1 circuit
2 Mbits	30 circuits
8 Mbits	120 circuits
34 Mbits	480 circuits
140 Mbits	1920 circuits
565 Mbits	7680 circuits

The characteristics of the digital circuit were defined by the International Telegraph and Telephone Consultative Committee (CCITT) and are known as the '700 Series' of recommendations. Amongst other areas they define circuit bit rates, interfaces, frame structure and fault conditions.

Basing their offerings on the above recommendations and standards, the two main UK carriers offer a number of digital transmission services. These are broadly grouped into 'switched' and 'non-switched' (point-to-point) services. The switched services basically comprise X.25 packet switching and the emerging ISDN. Both these services are covered in detail in Chapters 6 and 10 respectively. The non-switched services are covered here and can best be categorised by their transmission speeds, ie *kilobit* and *megabit*.

Kilobit Transmission

British Telecom's 'KiloStream' and Mercury's 64 Kbits service options offer a digital point-to-point leased circuit, operating synchronously at a number of data transfer rates up to 64 Kbits per second. Such a service provides the physical interconnection only. It is totally transparent to any protocol or format, thus allowing the subscriber complete freedom in the use of the service. In the case of BT's KiloStream the digital circuit can normally utilise existing four-wire cabling between the local exchange and subscriber premises, thus decreasing the lead time for circuit provision. Mercury also has a number of methods available for circuit provision, including the installation of microwave links.

The digital circuit from the local exchange is terminated at the subscriber premises by a circuit termination and customer interface unit British Telecom call the *Network Termination Unit* (NTU). The Mercury equivalent of the NTU is the *Circuit Termination Unit* (CTU). The NTU/CTU devices are mains powered units which provide an interface between a subscriber's communications equipment and the incoming digital circuit. In conceptual terms the NTU/CTU can best be considered the functional equivalent of a modem on analogue circuits.

Several variants of the unit are available, depending on which interface is required. These include:

— X.21 bis at selected speed ranges of up to 19.2 Kbits, giving V.24/V.28 (RS232C) compatibility;

— X.21 bis at 48 Kbits, giving V.35 compatibility;

— X.21 at selected speed ranges of up to 64 Kbits (at 64 Kbits the X.21 interface consists of X.24/X.27 data and timing elements only).

At data transfer rates of up to 48 Kbits, the NTU/CTU adds two supervisory bits to every 6 bits transmitted by the user, which enables the PTT to provide in-service monitoring and fault diagnosis between the local exchange and the NTU/CTU. This facility to monitor between the exchange and customer interface is termed a *structured service*. It is offered in addition to the extensive monitoring undertaken within the PTT's internal network structure.

At the full 64 Kbits rate, no in-service monitoring is available and fault diagnostics are at a reduced level. The diagnostics can only be initiated when the circuit is out of service to the subscriber, so subscriber co-operation is required. The advantage, however, is the increase in transmission speed. The service here is termed an *unstructured service*.

The tariffs for kilobit services consist of an initial connection charge and an annual rental; the rental is normally calculated on the circuit distance. The tariff also includes the provision of an NTU/CTU at each end of the circuit.

Megabit Transmission

British Telecom's 'MegaStream' and Mercury's megabit service options offer a range of high-speed digital circuits from 2 Mbits up to specially engineered circuits operating at 140 Mbits. The 2 Mbits option is currently the service most favoured by end users. As with kilobit transmission, it can be presented either as a structured or unstructured service, and is totally protocol-transparent. With the 2 Mbits structured service, the circuit is generally provided to the subscriber in the form of thirty 64-Kbits channels, although other options offering different combinations of channel numbers and channel speeds are available.

Where thirty 64-Kbits channels are supplied, the PTT uses the remaining bandwidth of the 2 Mbits circuit to provide two 64-Kbits channels for monitoring, diagnostic and synchronisation purposes. The 30 subscriber channels can be used to support voice or data traffic, or a combination of both. The channels can be routed on either a single point-to-point basis, or, as an alternative, one end of the circuit could terminate all 30 channels while each individual component channel could be terminated at up to 30 different geographical locations. With the addition of multiplexers, each 64-Kbits channel can be further subdivided into several lower speed channels as required.

The 2 Mbits unstructured service presents the whole bandwidth as a 2 Mbits 'pipe' to the subscriber, thus allowing very high-speed transmission between two attached devices. The disadvantage of this service is that only limited diagnostics and testing facilities are available from the PTT between the local exchange and the user premises.

With both the structured and unstructured services, additional diagnostic, monitoring and network management facilities are available to the subscriber with the introduction of additional equipment (such as multiplexers) at the subscriber termination ends. The interface presented to the subscriber generally conforms to the CCITT G.703 recommendation.

An early criticism of both KiloStream and MegaStream was the unreliability of the service, so from 1985 BT began to issue performance and reliability targets as a general estimate and guide to users of this service. With British Telecom's overall assessment of these services, it was judged that over a period of at least one year, the circuit availability would in most cases be better than 99.85%. A circuit is deemed to be unavailable when there is a break in transmission (no data throughput) for longer than 10 seconds, or when the error ratio exceeds 0.001% for more than 10 consecutive seconds.

PTT Digital Network Infrastructure

At this point it may be useful to look at the internal digital network infrastructure of a PTT such as British Telecom.

The BT Network Example

Let us take as an example a subscriber who requires a digital 64 Kbits transmission circuit between two remotely located premises. The PTT would first install an NTU device at each of the subscriber premises. The NTU would then be connected to the nearest local exchange by some form of physical connection such as four-wire cabling or optical cabling, or via a microwave link. Next, a physical connection is established across the internal network between the two exchanges. The connection is not one long physical cable, but a number of shorter physical sections connected together at various points across the network to form a logical channel for the subscriber.

The points where the sections of the circuit are connected together are termed cross-connection sites. They are centres where large numbers of circuits are brought together, terminated, and then physically interconnected to enable incoming channels to be onward routed through the site towards their final destination. The cross-connection process was originally a manual operation: an engineer physically linked the two incoming circuit termination points together to form the circuit. However, this manual approach was subject to problems, particularly when a circuit was being established across the network through several cross-connection sites. Co-ordination was required so that each site correctly connected its section of the circuit at the required place and time. Obviously this manual process was subject to errors, as well as costly in terms of manpower and time, so an automatic solution to cross-connection was sought.

The original specification for Automatic Cross-connection Equipment (ACE) called for an extremely flexible device, capable of terminating a number of 2 Mbits bearer circuits and demultiplexing them back down to their individual 64 Kbits channels. These would then be routed through the device according to a predefined route before being remultiplexed back onto an outgoing 2 Mbits bearer circuit. The device would carry out this function automatically using its own internal intelligent processors, but it would also have the capability of being remotely controlled from a central control centre, where network operators could configure and manage all ACEs within the network.

This centralised control function is an important feature, as it enables a circuit to be established through a number of ACE sites quickly and with the minimum of manual intervention. Once the physical circuit and the logical channel have been established, the subscriber can attach his communications equipment to the NTU interface. The NTU's function is to encode the outgoing traffic into either a structured or unstructured format before transmitting it to the local exchange. The signal is encoded by the NTU using a technique known as 'WAL2'.

At the local exchange, the incoming kilobit circuit is terminated at a device known as a *primary multiplex*. This is basically a sophisticated time division multiplexer capable of terminating up to 30 similar circuits each operating at speeds of up to 64 Kbits. In addition to the incoming circuits from various local locations, the primary multiplex has an outgoing 2 Mbits circuit which is the main bearer circuit into the heart of the PTT network. The primary multiplex's function is to combine all the incoming data streams onto the 2 Mbits bearer circuit using the time division technique. Because the 2 Mbits binary stream cannot be directly transmitted at this speed, it is converted and encoded into a format known as 'High Density Bipolar 3' (HDB3).

The cross-connection site acts as a large circuit switching point for a number of local exchanges. The subscriber's 64 Kbits data traffic is routed through the site according to its predefined circuit path. It is again multiplexed back onto a 2 Mbits bearer circuit to continue its journey across the network towards its eventual destination. On this journey, the 2 Mbits bearer circuit itself may be multiplexed up onto higher-speed bearer circuits.

The 'second-order muldex' implements the second step in the CCITT recommended hierarchy for digital networks (CCITT recommendation G.742). It accepts four 2 Mbits circuits and multiplexes them to form a composite 8 Mbits bearer circuit. The 'third-order muldex' carries out a similar function in accordance with CCITT recommendation G.751. Four 8 Mbits bearer circuits are multiplexed onto a 34 Mbits bearer circuit. This process continues, with four 34 Mbits circuits being multiplexed onto a 140 Mbits bearer circuit, and four 140 Mbits circuits onto a 565 Mbits bearer circuit. At this level, 7680 channels at 64 Kbits each are being combined over a single physical path. The subscriber's single 64 Kbits circuit could therefore pass through a number of multiplex/demultiplex stages and several cross-connection sites before finally arriving at the remote PTT exchange. There the 64 Kbits circuit passess outwards to its final destination: the subscriber's remote location and the NTU. The data stream is then decoded and transmitted via a standard interface to the target DTE device, (see Figure 2.1).

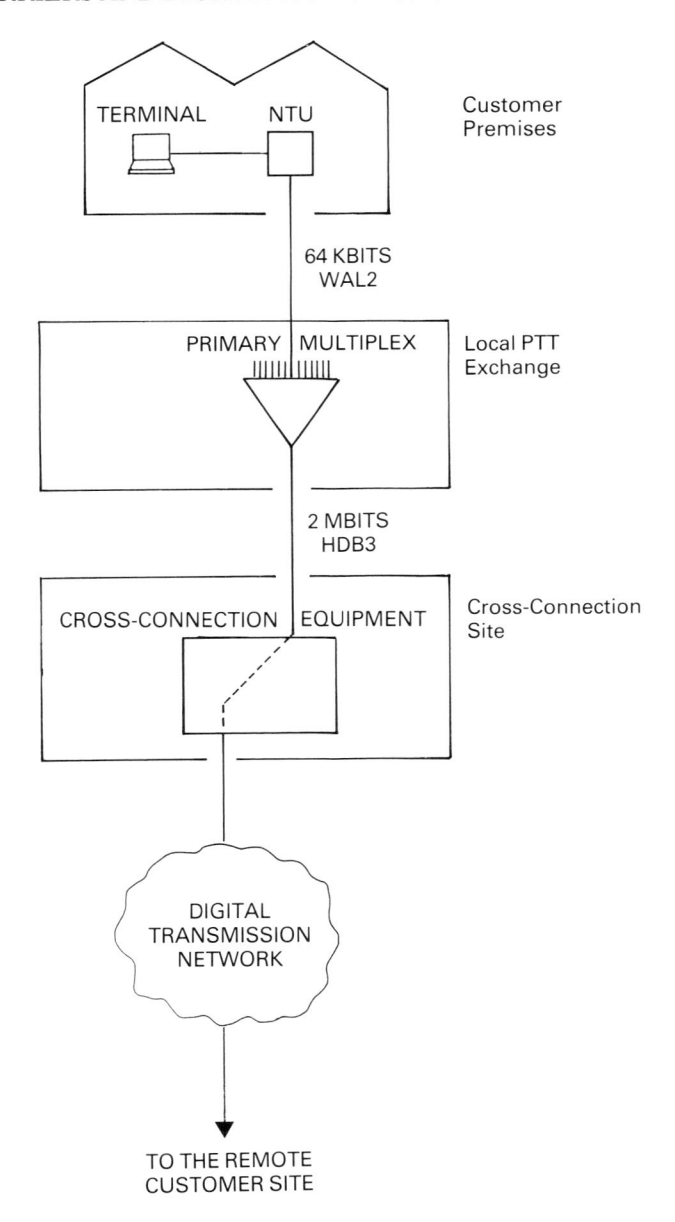

PTT CARRIERS AND DIGITAL TRANSMISSION SERVICES

Figure 2.1 PTT Digital Network Infrastructure

Some subscribers require multiple 64 Kbits circuits to be terminated at one location. In these cases the primary multiplex, normally located at the local exchange, is moved to the customer's premises. A 2 Mbits circuit forms the link between the primary multiplex and the local exchange. The primary multiplex at the subscriber's premises then allows up to 30 individual 64 Kbits circuits to be routed around the local area.

Megabit circuits operating at 2 Mbits or above are installed in a similar manner. The megabit circuit is brought direct from the local exchange and terminated at the customer's premises. A CCITT standard G.703 interface is then presented to customer equipment such as proprietary megabit multiplexers or digital private telephone exchanges.

North American Digital Services

At this point a quick overview of the United States digital network services is well worthwhile. This is partly because some UK organisations require connections to this part of the world, but mainly because the majority of digital equipment available in the UK was originally developed for the North American market.

Since its introduction in 1974, AT&T's 'Dataphone Digital Service' (DDS) has become a fairly well established communication medium. Functionally equivalent to British Telecom's 64 Kbits service KiloStream, DDS provides leased digital circuits with a maximum transmission rate of 56 Kbits, although, like KiloStream, lower speeds are available. DDS has connections to BT's KiloStream service and Mercury's kilobit services, as well as connections to other equivalent digital networks such as Canada's 'Dataroute'. The DDS equivalent to the NTU/CTU is made up of two devices, the *Data Services Unit* (DSU) and the *Channel Services Unit* (CSU), although the trend is to integrate these two devices into one unit.

'Accunet Switched 56' from AT&T is a 'pay for what you use', dial-up 56 Kbits digital data service available in the US and connected to a similar service in Canada.

The T-Carrier system is the North American PTT industry standard for Megabit digital communications. These range from 'T1', which operates at 1.544 Mbits, through to 'T4', which operates at 274.176 Mbits. 'Accunet T1.5' is a T1, leased-line service from AT&T for point-to-point digital communications. This service can be considered the functional

PTT CARRIERS AND DIGITAL TRANSMISSION SERVICES

equivalent of the 2 Mbits BT MegaStream or Mercury megabit services in the UK. 'Accunet Reserved 1.5' offers point-to-point T1 links on a scheduled basis. The user schedules the service in advance with AT&T and pays for it in one hour increments, thus reducing costs as compared with leasing a full-time dedicated link.

USER ISSUES REGARDING DIGITAL SERVICES
The Benefits of Digital Communications

There are a whole host of benefits and cost savings available to the PTTs through the installation and use of digital technology within their networks. The big question for most of their customers, however, is what's in it for the digital subscriber now and in the future. Clearly there are disadvanges as well as advantages, and the answer varies according to the customer's current needs and future requirements. The following summarised points may go part way to answering the question, and again mainly relate to the digital non-switched services from the PTTs.

Cost Savings

The analogue circuit still tends to have a cost advantage over the digital equivalent, especially in terms of initial installation charges and ongoing rental charges for circuits over a relatively short distance. However, this cost advantage on distance is reducing as time goes on. The break-even point on a 9.6 Kbits digital or analogue circuit is now around the 270 km mark; the greater the distance, the more cost-effective digital usage becomes. This calculation does not take into account the fact that modems would be required for the analogue circuit, whereas the NTU/CTU is included in the cost of the digital circuit rental.

Transmission Quality

Analogue circuits have always been subject to circuit noise and distortion, which slow the overall throughput capacity of an affected circuit. Digital transmission does not suffer from similar problems and is the ideal medium, especially over the longer distances. Although improved transmission quality benefits traditional voice traffic on a corporate digital network, it is particularly relevant to the corporate data network, where improved circuit quality decreases the potential for transmission errors and failures and thus improves overall throughput.

Transmission Capacity

Digital circuits have a wide bandwidth capacity in comparison with most analogue circuits. In particular, megabit circuits can be subdivided into an almost infinite number of channel and speed combinations, allowing large numbers of data, voice, text and image channels to be routed over a single digital circuit.

Transmission Speed

An alternative view of the digital circuit bandwidth is to see it as a single high-speed channel. This gives the end user a very fast link which can be applied to a number of applications. For example, remotely sited mainframe computers can undertake practical bulk file transfer tasks; it could enable channel-to-channel attachment of remotely sited processors; or it could be used to correct remotely located PABXs or similar equipment.

Integration

Integration of voice, data, text and image communications onto a single corporate network is possible through the use of digital circuits and other equipment such as intelligent multiplexers, switches etc. Digital circuits could also provide the physical infrastructure for a private ISDN. (Integration is covered in more detail in the following chapters.)

Circuit Provision

From the PTT's viewpoint, it is easier to provide an individual digital circuit than the corresponding analogue circuit. This is because all circuits are made up of a number of individual sections passing through automatic switching points, each section being a small proportion of the internal wideband trunk circuits. The digital network has the advantage of modern automation in setting up a circuit, reducing both the time and the manpower coordination needed to achieve this. In British Telecom's case, use can also be made of existing four-wire telephone circuits from the local exchange to the customer termination point when installing KiloStream circuits. These factors allow installation time and costs to be reduced, which benefits not only the PTT but the digital subscriber as well.

PTT CARRIERS AND DIGITAL TRANSMISSION SERVICES

Additional Points

Other benefits of digital transmission include improved fault and error monitoring facilities for both the supplier and customer on digital networks; reduced call set-up times, so that links can be established over a private digital switched network in a fraction of the time taken by conventional analogue equivalents; and the use of high-speed encryption facilities that do not affect general throughput speeds.

Like all technology, however, digital communication does have drawbacks and disadvantages. One possible disadvantage is the use of digital bandwith when compared with bandwith usage on analogue circuits. On speech telephony, for example, digital circuits tend to use a bandwidth of up to 64 Kbits. Analogue speech can be achieved with a bandwidth of less than 3 KHz (equivalent to around 10 Kbits). This disadvantage is offset to a large extent by the overall bandwidth available on a digital circuit. The cost of a digital application such as voice transmission is falling rapidly in comparison with its analogue equivalent.

Another disadvantage of digital transmission is the problems that can occur if a digital link fails. In an analogue network, the failure of a circuit would probably cause disruption to a small number of end users, and even then there might well be the option of a dial back-up standby service over the PSTN.

When a digital link fails everybody knows about it, mainly because the digital circuit tends by its nature to carry far more traffic of various types, including voice and data. If the failure occurs on a main megabit link in a corporate network, it could be potentially disastrous to the entire user population.

It is therefore important to consider how such failures can be bypassed. The current approach to network resilience relies mainly on careful network design. As much alternative routeing capability is built in as practicalities and money allow.

Subscribers who require corporate network resilience often consider duplicating megabit circuits between two locations, or providing alternative routes via intermediate locations. However, care should be taken to ensure that such circuits are not routed through the same local PTT exchange, as failure here could result in the loss of all circuits. At additional cost, the PTT can usually arrange to route circuits from a customer's premises to two or more exchanges, thus spreading the failure risk. Alternatively, some organisations obtain megabit circuits from more than one PTT, thus providing completely separate routes for their corporate communications.

Resilience for kilobit circuits can be obtained by similar methods as those described for megabit circuits, but in addition kilobit back-up facilities can also be provided by the switched digital services: packet switching and the emerging ISDN. Obviously packet switching is limited to providing data services only, whereas ISDN has great potential for providing a flexible public dial back-up service for a private corporate network.

3 Digital Multiplexing

GENERAL BACKGROUND TO DIGITAL MULTIPLEXING

Multiplexing is a technique that permits a single physical communications circuit to support across it a number of logical communications channels. Digital multiplexing techniques were originally developed in the early 1960s with the introduction of 'Pulse Code Modulation' (PCM) systems on the inter-exchange connections of the PTT networks. This type of technology was soon in demand by the end users, particularly the data communications fraternity, who by this time had a growing population of data processing equipment which increasingly relied on simple point-to-point analogue circuits for their basic communications needs. They also had a problem, in that demand for such circuits was rapidly outgrowing supply. This in turn resulted in escalating communication costs to their user organisations.

The data multiplexer was conceived and developed to provide better utilisation of existing circuits. This development solved the problems of circuit availability and cost by reducing the need for multiple parallel circuits between two locations, whilst making cost effective use of the remainder.

But for these earlier experiences the problems encountered with analogue data circuits could have recurred with the introduction of digital point-to-point circuits. The digital multiplexer initially followed the principles established by its analogue counterpart. It soon evolved into a more complex device as the need for advanced data transmission applications and the incorporation of voice, text and image over digital circuits became a requirement.

The communications equipment manufacturers have today brought a whole range of digital multiplexers to the market place. These range from the basic point-to-point device supporting a few channels over a 64 Kbits circuit to advanced multiplexers designed to interconnect with other similar devices to provide a distributed, resilient, intelligent switching network. This can reflect the communications needs of an organisation's distributed and complex structure.

The basic multiplexer normally enables the network designer to subdivide the digital circuit's bandwith into a number of different combinations, giving several low-speed channels or a combination of low-speed data channels and one or more voice channels (see Figure 3.1). At this basic product level, the time division multiplexing devices normally only support synchronous communication, although asynchronous support is increasingly being provided by the incorporation of statistical multiplexing techniques into the basic device. Basic multiplexers will normally incorporate simple test functions which are accessed via external switches, usually located at the front of the device. The results of tests are indicated on LED indicators; as a rule, these are also located at the front. Configuration, similarly, is carried out by setting switches or multi-position dials either at the front of the device or internally.

The basic multiplexer is a relatively cheap way to improve the throughput of a digital kilobit circuit at a relatively low cost. The only disadvantage of the basic device is its limited ability to integrate into an overall network management system: it cannot generally be upgraded as the network grows. Both these problems can be circumvented to a large

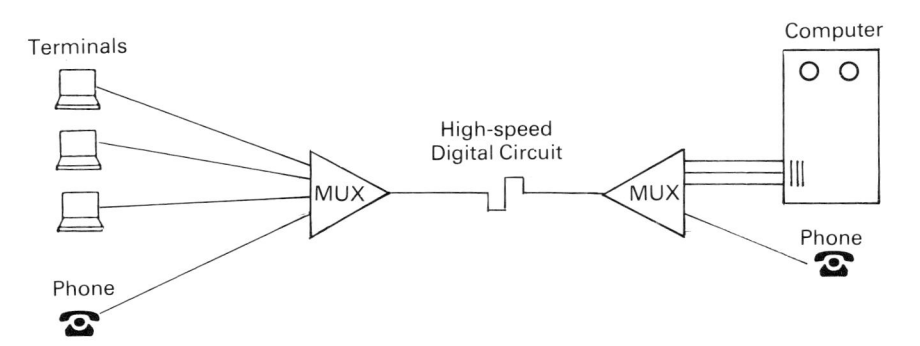

Figure 3.1 Multiplexers Utilise the Digital Circuit to its Maximum Potential

DIGITAL MULTIPLEXING

extent by installing additional stand-alone equipment such as 'wrap-around' network management systems, circuit or channel switches, or a number of other communications devices which improve the performance and flexibility of the network whilst still incorporating the original basic multiplexer components.

The more advanced multiplexer is generally an 'up-market' version of the basic device, giving a number of additional options and features to the network designer. In most cases this type of multiplexer supports a number of high-speed digital circuits, enabling a flexible network structure to be created by the network designer. For example, to provide circuit resilience on important communication links two digital circuits could be installed in parallel, the circuits operating either in a 'load sharing' mode (with the traffic being shared between the two) or in a 'primary and secondary' mode (with the latter providing a back-up alternative route for the primary circuit in the event of its failure). In both cases the attached multiplexers would provide the intelligence for either the load-sharing or circuit-rerouteing operations.

A more cost effective approach to this scenario would be to construct a network of multiplexers with channel routeing capabilities. In such a network, incoming channels can be internally routed through the multiplexer back on to other outgoing channels. (This routeing capability is normally undertaken without demultiplexing and remultiplexing the signal, an important factor if the channel is carrying voice traffic.) Consider a network of three or more such devices, each with a single circuit link to its neighbour. If a direct link between two multiplexers fails, an alternative route could be established automatically through one or more of their interlinking neighbours. This provides the network with resilience in the event of circuit or multiplexer failure (see Figure 3.2).

These types of software programmable multiplexer usually allow for supervision and control to be carried out by an attached supervisory asynchronous terminal, PC or similar device. Many have a supervisory RS232 interface port to allow direct connection of these devices. Remote supervision could be established over the PSTN or PSS, assuming the availability of modems and some form of error-protected supervisory protocol. Alternatively remote supervision may be established over the existing corporate network using spare bandwith on the digital circuits. Either way, the supervisory terminal should allow the network operator to configure, test and control any multiplexer on the network, be it

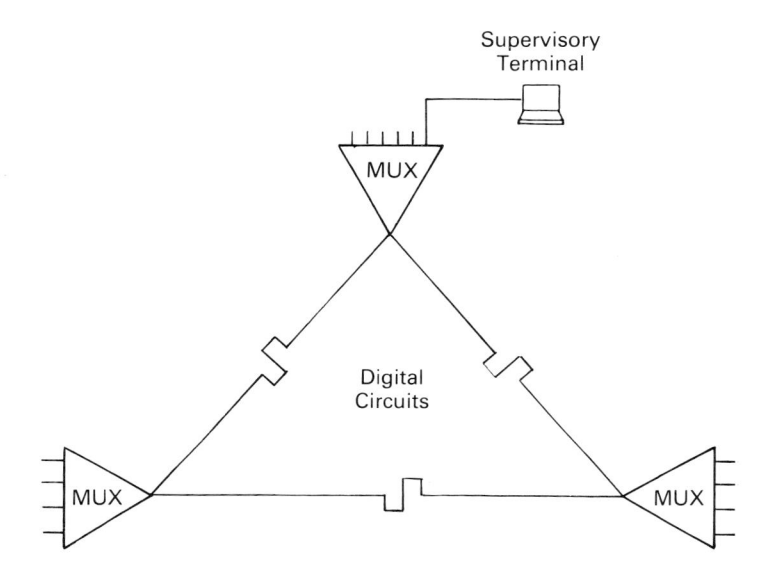

Figure 3.2 Advanced Multiplexers Have Channel Routeing Capabilities

directly attached or remotely sited. The only limitation on such a set-up is that the multiplexers and supervisory software are from the same vendor.

Some multiplexers provide a secondary interface port for the attachment of a printer. This allows a hard copy printout to be made of multiplexer and network activity, diagnostic tests and statistics. The secondary port may also be used for attaching a second supervisory terminal.

Advanced multiplexers can normally support combinations of synchronous, asynchronous and isochronous data traffic at a variety of speeds, as well as voice channels interfacing to handsets or directly into digital telephone exchanges. Telex, telegraph and similar low-speed data traffic, typically operating at speeds of only 300 bps or less, can be handled economically across a digital network by using a multiplexer to concentrate these channels onto a single 64 Kbits circuit.

Multiplexers are available which allow the direct connection of megabit circuits supporting a large number of channels. The more advanced kilobit multiplexers usually have the option to be field upgradable to

DIGITAL MULTIPLEXING

support megabit operation, making such devices fairly flexible in their use. Because of the characteristics of megabit circuits, megabit multiplexers today tend to only use time division multiplexing techniques. In the selection of any multiplexer, consideration must be given to its transmission efficiency, throughput delay, protocol handling abilities, manual and automatic channel routeing capabilities, buffer management methodology and diagnostic facilities. Several alternatives exist for each of these factors; their use or even their existence in a particular multiplexer model largely depends on the design objectives of the manufacturer. These in turn are based on their perceived view of the market requirements and needs. Above all these selection criteria, however, the network designer must regard the resilience (or uptime) and device/network manageability as the most important aspects affecting digital multiplexer selection.

TECHNICAL CONCEPTS BEHIND MULTIPLEXING

For those readers who are not familiar with multiplexing concepts, this section gives a brief overview of the various ideas behind the devices on offer today, together with some of the advantages and disadvantages gained by the various methods. The basic multiplexing techniques discussed are those related to digital transmission, ie time division multiplexing and statistical time division multiplexing.

Both these techniques are available in a wide range of models ranging from a basic device for simple point-to-point links through to advanced networking devices capable of linking numerous digital circuits into a complex network. Within this wide range of models, a number of options and facilities are built in. (Devices of more modular construction may have the capability of being field upgraded to user requirements.) Some multiplexers incorporate many of the facilities of digital switches, and use techniques relating to network control and management: these areas are discussed in Chapters 4 and 8 respectively. This chapter will concentrate on features which are more or less fundamental to multiplexer design and use.

Time Division Multiplexing

As the name suggests, Time Division Multiplexing (TDM) is a technique based on allocating time slots of a fixed duration for each input low-speed channel logically connected to the multiplexer. These individual time slots are assembled into a 'frame' to form a single data block in a high-speed data stream. The building or interleaving of the frame takes two forms, *bit-interleaving* and *character* or *byte-interleaving*. For a number of reasons bit interleaving is more commonly used on digital multiplexers: Figure 3.3 shows a simplified representation of the bit-interleaving process.

The data from the device attached to the input low-speed channel is received by the multiplexer in a bit format. A high-speed scanner inspects each channel in turn, and (assuming a data bit is present) transfers it to the output frame data stream. If no data is present an idle indicator bit is placed in the output frame. By the time a complete scan of all the input channels has been made and the frame created, a unique bit sequence known as a 'frame synchronisation pattern' or 'framing pattern' has been added. This provides synchronisation between the transmitting and receiving high-speed scanners, thus ensuring that the correct input channel bit arrives at the correct output channel. The framing pattern in some designs of multiplexer is distributed through a number of frames rather than a single frame so as to increase the overall data throughput.

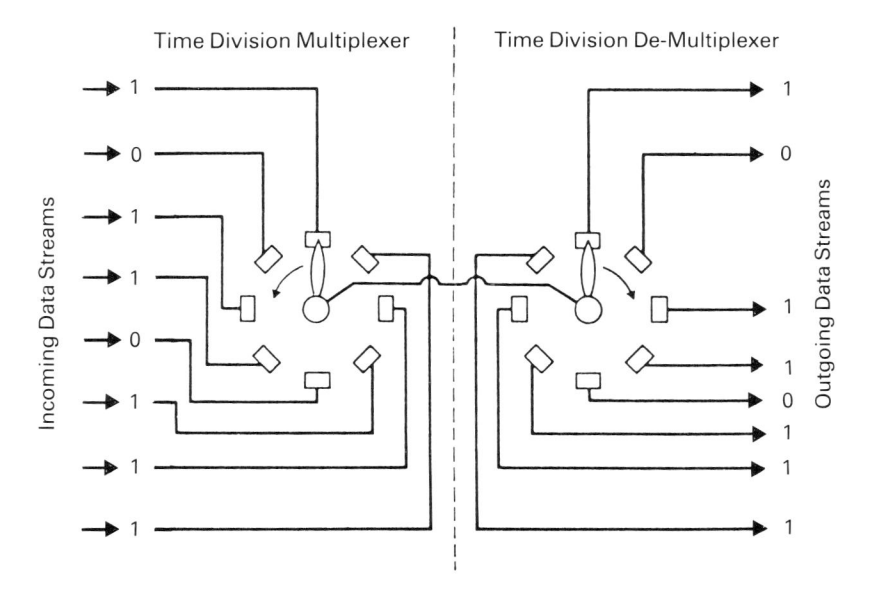

Figure 3.3   Principles of Bit-interleaved Time Division Multiplexing

DIGITAL MULTIPLEXING 37

At the receiving multiplexer, the demultiplexing stage must detect the framing pattern a predetermined number of times before synchronisation is assumed. If synchronisation is lost for any reason, re-synchronisation is undertaken, and is not assumed until the framing pattern has again been detected the predetermined number of times.

Character-interleaving works by a similar method to bit-interleaving, except that characters instead of bits are placed in the frame. (See Figure 3.4.) Bits are received from the attached devices by the multiplexer and placed in a buffer known as a *serialiser*. Once a complete character from a low-speed channel has been received, the contents of the serialiser are moved to a buffer register. A high-speed scanner then inspects each channel's buffer register: if a character is present, it is transferred to the

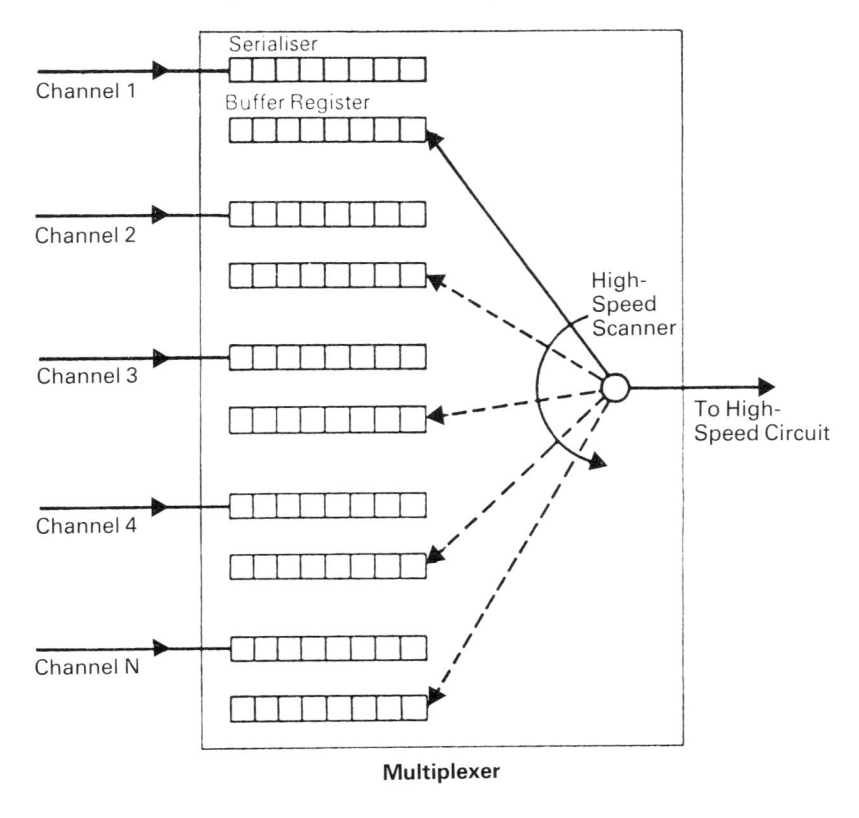

Figure 3.4 Simplified Character-interleaved Time Division Multiplexer

output frame data stream. As with bit-interleaving, if no data is present an idle character is placed in the output frame.

Frame synchronisation and frame control information are added to the data stream for multiplexing control functions. However, because characters rather than bits are interleaved in a frame, the data stream frames tend to be longer, which results in a greater ratio of information bits to synchronisation bits. The greater efficiency which this allows is offset to some extent by the longer time taken to re-synchronise when errors occur.

Bit versus Byte Interleaving

Bit-interleaving is generally used with multiplexers operating synchronously, as synchronous communication is by definition bit-oriented in its transmission method. Character-interleaving is more oriented towards asynchronous devices using a character-based protocol. As buffering is used for character-interleaving multiplexers, the designers can apply a 'stripping' technique to the incoming asynchronous data stream. The start and stop bits that surround the data characters are removed during the multiplexing stage and reinserted at the demultiplexing stage, thus improving throughput over the circuit by eliminating unnecessary bits. Character-interleaving also allows greater efficiency in the bypass of channels in a multi-node multiplexer network.

As previously mentioned, character-interleaving normally provides a better 'data-to-synchronisation bit' ratio than bit-interleaving techniques. This throughput advantage is offset to some extent by the additional time taken during re-synchronisation, as well as the time taken by internal buffer processing delays. Character-interleaving multiplexers are generally more complex in design and use significant amounts of RAM for buffer storage, which as a result adds to the price of the device. However, recent techniques which dynamically allocate buffer space to active channels have tended to reduce this RAM requirement.

The Constraints of Fixed Duration TDM

There are disadvantages associated with the 'fixed duration' or 'rotational' time division multiplexing techniques described above. Chief amongst these is that the scanner examines every channel in turn for data, inserting idle indicators into the frame when no data is present. This can be viewed as inefficient use both of the multiplexer and of the high-speed

DIGITAL MULTIPLEXING

circuit, especially when in theory the actual data could be transferred instead of 'idle' control characters. The circuit has in effect been slowed by the inter-multiplexer control protocol. This disadvantage could also be viewed as a constraint of this type of TDM operation. The sum total of the input channels cannot exceed the speed of the high-speed digital link. Therefore to run four channels at 2.4 Kbits, the digital link must be operating at a speed of at least 9.6 Kbits.

Dynamic Time Division Multiplexing

Dynamic Time Division Multiplexing extends the concept of the fixed duration TDMs by using a technique known as *dynamic bandwidth allocation*. Conventional TDM techniques share high-speed digital circuit capacity in a fixed manner, with each low-speed channel being allocated a time slot whether or not it currently requires it. A dynamic TDM has the ability to monitor each of the low-speed channels, allocating time slots to these channels only when such a channel has data to transmit. In effect, a predefined portion of the high-speed circuit is allocated to the low-speed channel, the two usually being equivalent in terms of bandwidth or speed. This allows the maximum use to be made of the high-speed digital circuit, as well as providing the optimum throughput speed for connected users. However, if the high-speed digital circuit is operating at full capacity a 'contention' formula is applied, and additional users who wish to obtain a channel through the multiplexer are queued on a user priority basis. The aggregate speeds of the currently active low-speed channels cannot exceed that of the high-speed circuit.

The Pros and Cons of TDM

Time division techniques are generally 'protocol transparent' (eg the TDM can handle any protocol type), so they are particularly appropriate to applications where the data is in a new, obscure or non-conforming protocol format. The only prerequisite in these cases is that the TDM can recognise the beginning and end of the information. This is achieved either by control signals or by predefining data headers and trailers.

The ability to intermix data and voice is one of the major benefits of digital TDM multiplexers. It enables an organisation to obtain the maximum return from its investment in a digital network, and is used by many network designers as a major justification for switching from analogue to digital networking. This feature works by taking circuits from individual telephone handsets or from PABX telephone exchanges and linking them into the low-speed channel interfaces of a local multiplexer. This device then encodes and multiplexes the voice channels onto outgoing digital circuits. At the remote multiplexer, voice channels are routed out to the appropriate user location.

The voice encoding methods available on today's multiplexers are a playoff between quality of transmission and bandwidth used. Early methods used a bandwidth of 64 Kbits, and one of the first to become an internationally accepted standard was *Pulse Code Modulation* (PCM). The basis of PCM is to sample an analogue speech signal at regular intervals, convert this sample into a series of digital pulses or codes, and then reverse the process at the receiving end to recover the signal.

Early PCM systems sampled the analogue speech signal 8000 times a second. The sample was then converted into a seven-bit code, together with an eighth bit to signify the sign of the signal; this constituted an *octet*. The bit rate of a single PCM channel was thus 8 bits × 8000 samples a second, equalling 64 Kbits per second transmission. The standard telephony PCM systems used TDM techniques to carry 32 channels consisting of 30 traffic channels, one synchronisation channel and one signalling channel over a 2 Mbits circuit. As PCM was developed, lower bandwidth usage algorithms were devised, operating at speeds as low as 32 Kbits. The reduction in transmission speed rate is achieved by reducing the number of samples per second. This results in a reduction of transmission quality, but at the same time allows multiple voice channels or a combination of voice and data channels over a single 64 Kbits digital circuit. Alongside the development of PCM, other voice encoding methods were being introduced.

Continuously Variable Delta Modulation (CVSD) was another voice encoding method which was developed to use a variety of speeds up to 64 Kbits. CVSD samples the incoming speech signal in a similar manner to PCM, but rather than converting it into a bit code, it generates a series of '0's and '1's. Thus when the signal frequency rises '1's are transmitted across the digital circuit, and when the signal frequency falls '0's are transmitted. Altering the rate at which the speech signal is sampled increases or reduces the generation rate of the '0's or '1's, which in turn affects the transmission speed requirement across the digital circuit. Like PCM, the transmission speed of the voice channel reflects the resulting quality obtained.

DIGITAL MULTIPLEXING

Adaptive Differential Pulse Code Modulation (ADPCM) is one of the latest methods of voice encoding to be developed, and is expected to become the prime international standard on both the private and public networks of the future. Voice channel bandwith speeds of 4.8 Kbits have been obtained through the use of ADPCM, although at such speeds the resulting output quality is poor. ADPCM operates in a similar sampling manner to PCM. The main difference is that it only transmits to the receiving end when a change occurs in the incoming speech signal. For example, assuming a static signal frequency over a minute period of time, ADPCM only transmits at the beginning and end of the static signal, when the signal frequency changes up or down. (The signal samples in between, which are all static, are not transmitted over the digital circuit.) This approach obviously reduces the overall bandwith necessary to transmit a variable speech signal, but would have little impact on a fairly consistent frequency such as a carrier signal.

Basic asynchronous data terminal devices are a problem area for all TDMs. Because of the lack of error checking and correction built into such user equipment, and the complexities of integrating asynchronous data into a synchronous transmission link, TDMs incorporating both asynchronous and synchronous support tend to be more complex and costly. An alternative approach to asynchronous support is to place a statistical multiplexer between asynchronous devices and the TDM.

Statistical Time Division Multiplexing

Statistical Time Division Multiplexing (STDM) works on a 'transmit or store and forward' basis. This principle works on the assumption that incoming low-speed channel data is assembled into a frame and transmitted as and when available. During periods when the high-speed circuit is fully utilised, the incoming data will be temporarily stored until bandwith becomes available again.

Where the low-speed channels attached to the multiplexer are lightly utilised, it is theoretically possible for the total aggregate data rate of these channels to far exceed the high-speed digital circuit's speed. In practice some form of buffering function will occur if the actual total aggregate data rate of the low-speed channels exceeds that of the digital circuit. STDMs using this multiplexing method therefore contain large buffers capable of absorbing short-term peaks in demand. In addition, to prevent buffer overflow and loss of data, a mechanism is normally available to suspend data flow into the multiplexers. This *flow control* takes two forms: it will either output 'X-OFF' control characters to the transmitting device via the 'receive data' circuit in the terminal interface, or drop the 'clear to send' signal in the terminal interface. Once the multiplexer is able to resume these channels, either the 'X-ON' control character is transmitted or the 'clear to send' signal is raised.

An important consideration with flow control techniques is the amount of data transmitted by the attached device after the 'suspend transmission' signal has been sent. Some devices may transmit a few hundred characters before suspending, so it is crucial to match such devices with multiplexer reserve buffer capacity capable of handling such variations.

This type of statistical multiplexing technique should not suggest that bigger buffers are necessarily better. For maximum efficiency very little buffering should occur. Data should pass through the STDM with little or no delay, and buffering should only be necessary to absorb the data transmission peaks. A reasonable buffer size is around six to eight times the high-speed circuit's capacity, assuming that the buffer management design works on the principle of dynamic buffer management. More buffering would be required if buffers are allocated in a predetermined fashion during the initial STDM configuration stage. Buffer capacity above ten times the high-speed circuit's capacity is usually wasted and could even be detrimental: it can increase delay times, as well as increasing STDM costs.

When considering buffer management, thought should also be given to the time-out values preset within the attached user devices. Lengthy delays within the STDM could result in such devices aborting communication links unnecessarily. Thus when addressing the technical considerations behind flow control techniques, it is also important to consider what effects flow control has on traffic throughput, and ultimately on the overall efficiency of the STDM.

Flow control is just one aspect of the total overhead imposed by any proprietary STDM system. Other overheads include the construction and format of the STDM frame; like TDM frames, this contains both user data and frame control information. The user data within an STDM frame is assembled from the incoming data streams of the attached low-speed channels. The frame control information is then added. This generally comprises terminal address information relating to each user

DIGITAL MULTIPLEXING

data field held within the frame; inter-multiplexer control instructions; and the frame error protection checks that ensure the accurate transmission and reception of frames.

The most common of these error protection methods are the *Cyclic Redundancy Check* (CRC) and *Automatic Retransmission Request* (ARQ). In simplified terms, frames to be transmitted are stored in the ARQ buffer while two CRC control bytes are added. The frame is then transmitted and the receiving STDM checks the frame, sending an 'ACK' if the frame was correctly received, or a 'NAK' if the frame was corrupted. In the latter case the frame is then retransmitted.

STDM Frame Generation

Many different techniques exist for generating an STDM frame. In the simplest method, a high-speed scanner sequentially inspects each of the incoming low-speed channels from the attached user devices. When data is present on a particular channel, it is moved to the frame buffer and transmitted. Each frame therefore equates to a single device's data block. This approach is adequate on channels with low traffic usage, but on busy channels it would cause unacceptable response time delays (see Figure 3.5).

A second frame generation technique allows data from many user devices to exist within a single frame. Each active user device is allocated a variable length data field in the frame, together with a corresponding address and 'number of data characters' control field. This technique has a higher transmission efficiency than the first method, but still suffers from decreasing efficiency as traffic usage increases. Both methods are termed *byte count* techniques, because they employ variable length data fields and field size indicators.

A third method allocates each attached user device a two- to six-bit field within the frame. User data is placed in the appropriate field, if available; otherwise an idle indicator is used. This method is thus most efficient at high traffic usage, but decreases in efficiency as traffic usage decreases. Being a bit-oriented method, it can easily accommodate data compression techniques.

The fourth method has a performance somewhere between that of the second and third methods described above. It makes use of *address indexing*. Instead of assigning a 1- or 2-byte address field to every active

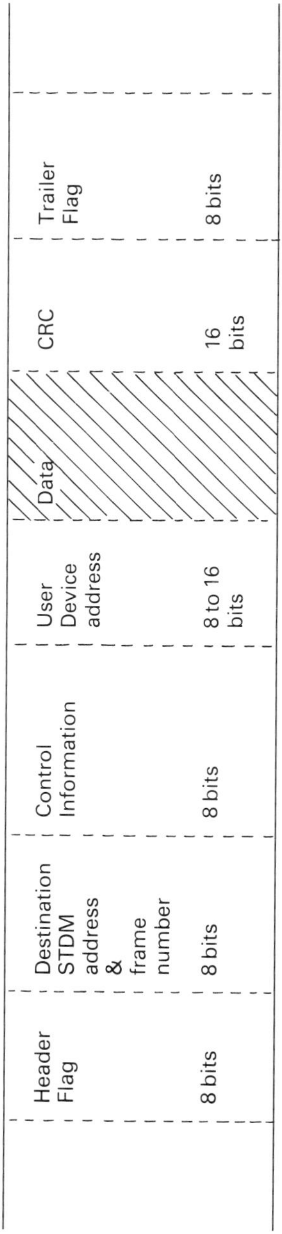

Figure 3.5 Simplified Example of an STDM Frame Layout

Header Flag	Destination STDM address & frame number	Control Information	User Device address	Data	CRC	Trailer Flag
8 bits	8 bits	8 bits	8 to 16 bits		16 bits	8 bits

user device, it uses a 2- to 4-bit address field to signify the relative position of the current data field's address in relation to the preceding data field's address. This method relies on the fact that in a high traffic usage environment there is a reasonable possibility that active user devices will be within the addressing range of the indexing field. For instance, a 4-bit indexing field can address from between 2 to 16 positions from its current address location. Differences in frame structure design can easily create a difference of between 5 and 15 percent in the transmission efficiency of otherwise comparable STDM products.

Data Compression

To gain maximum throughput on any given digital channel, a data compression technique should be employed by the STDM. For asynchronous data, it is general practice for the transmitting multiplexer to remove the start and stop bits from the incoming bit stream of an attached asynchronous device, and for the receiving multiplexer to replace these control bits at the demultiplexing stage. This 'bit-stripping' technique allows a pure data stream to be transmitted across the interlinking channel. On an 8-bit character protocol this gives a potential throughput improvement of up to 25 percent (2 control bits being removed in every 10 bits received by the STDM). In a similar way, BSC-type synchronous protocols can be subjected to a stripping technique where sync and pad protocol control characters are removed at the multiplexing stage and replaced at the demultiplexing stage.

Huffman encoding is an optimum compression technique for minimising the average character length. This procedure converts an 8-bit character (for example) into a variable length byte ranging between 3 and 7 bits. The most frequently used characters are converted to the shortest bytes, while the less frequent are converted to the longest bytes. The result is a 30 percent reduction in the average character length. Another compression method is to compact multiple occurrences of the same character, transmitting just the first occurrence of the character and an 'occurs' indicator. In this way, a string of 100 zeros can be compressed to one zero and an indicator stating 'occurs 100 times'.

The use of the latest compression techniques within STDM devices means that a data transfer which would normally have required a transmission speed of 14.4 Kbits can now be achieved in the same defined time, but operating at speeds of only 9.6 Kbits.

The Drawbacks of STDM

There are a number of issues to be considered by the corporate network designer in relation to digital multiplexer selection, acquisition and installation.

In some people's opinion the STDM technique now has the upper hand in comparison with TDM when used within a digital kilobit data communications environment. This is because of the superior facilities within STDM, such as advanced error protection and bandwidth utilisation. Design, however, is all-important in STDM devices. Differences in proprietary architectural concepts among the STDM vendors result in 25 to 50 percent differences in circuit utilisation efficiency alone.

All digital multiplexing techniques introduce varying degrees of delay to any communications link. Typically, TDM character-interleaved multiplexing adds between two to three 'character times' of delay to a synchronous data stream. Statistical multiplexing techniques, however, introduce even longer delays; this is due to the complexities behind the generation of STDM frames, error checking and data buffering features. Such delays can be kept to a minimum by allowing the network operator to 'fine tune' such features through the STDM control software interface, altering parameters such as frame size, buffer allocation and user priorities. Even so, STDM delays typically vary from 20 to 100 milliseconds per multiplexer. The delay factor is dependent on the traffic loading and processing power of the device, as well as other factors previously discussed.

Such delay factors virtually rule out the use of STDMs in any kind of corporate voice communication network. This is because delays or breaks in any conversation would prove unacceptable to the users concerned. The delay factor is also one of the reasons why telecommunications equipment manufacturers do not manufacture megabit STDM devices. One area where such delays would not present such a problem is in packet switching data networks. Statistical multiplexing is increasingly being used in this area, and is a major component of 'packet assemblers/ disassemblers' (PADs). (The use of statistical multiplexing techniques in a digital packet switching environment is discussed more fully in Chapter 6, Packet Switching and the Digital Connection.)

The methodology behind handling data protocols can significantly affect the performance of any protocol-sensitive STDM. Supporting

DIGITAL MULTIPLEXING

asynchronous formats is not particularly difficult, but statistically multiplexing synchronous protocols through a protocol-sensitive STDM can entail the recognition and actioning of up to 50 different protocol syntax control messages per protocol. This makes synchronous protocol handling a complex task requiring a thorough understanding of the different protocols by the STDM manufacturer.

Features of both TDM and STDM

Multiplexer Configuration

The configuration of any digital multiplexer from the most basic to the most advanced is conceptually identical. The main differences concern the ways in which the configuration settings are actually carried out. On the basic multiplexer, configuration is normally achieved by referring to the multiplexer instruction manual and physically setting some type of switch or multi-position dial on the device. Parameters such as transmission speed and protocol format are preset by the network operator on both the local and remote multiplexers. The number and combination of parameters are dependent on the device type, but (as always) the greater the number of parameter options available, the greater the chance of configuration errors. This is a definite disadvantage if the remote multiplexer is any distance away, as both time and manpower are needed to travel out and reconfigure it.

The advantage of the more advanced multiplexers is their 'software programmable' configuration. As well as or instead of physical switches, the local multiplexer will have some form of network operator interface that will allow the person to interact with inbuilt configuration software. This will not only enable the configuration of the local multiplexer to take place, but also permit the configuration of similar remote devices. From a single location, the network operator could (if necessary) configure and alter the entire multiplexer network.

Transparent versus Sensitive Multiplexers

Protocol transparency is a feature generally available on most TDM multiplexers today. It allows any synchronous protocol or asynchronous format to pass through the multiplexer. As such it allows the network designer great flexibility in planning for future network growth and developments, with the knowledge that the multiplexer can support any protocols required.

As an alternative to protocol transparent multiplexers, protocol-sensitive multiplexers are used where there is a requirement for the device to detect which protocol is being used. This is usually necessary where some form of conversion is required, so that the device can recognise and manipulate the incoming data stream correctly. This could, for instance, be used where a terminal can connect through the multiplexer to one or more incompatible computers in addition to its normal host. The multiplexer can then provide channel switching and protocol conversion where necessary. Another application for protocol-sensitive multiplexers is in the data compression function discussed earlier. To enable most data compression techniques to be applied, the multiplexer must as a prerequisite be able to detect the incoming data stream. The 'Automatic Baud Rate' feature may also be employed on protocol-sensitive multiplexers, enabling the device to detect the incoming speed, the outgoing speed requirement, and any conversion routine required.

Although protocol-sensitive multiplexers can be very useful in a network, the most important criterion for selecting such a device is that the protocols it supports now will match both your current and your projected future requirements.

USER ISSUES REGARDING DIGITAL MULTIPLEXERS

As digital multiplexers become more complex, their ability to handle a wider range of potential applications will become self-apparent. Below are a few of the facilities that could be incorporated into a digital network by the use of digital multiplexers, as well as a few of the potential pitfalls regarding digital multiplexer use.

Integrated Communications

The ability to integrate data, voice, text and image channels onto a single high-speed digital circuit must be one of the major benefits of digital communications. In this type of integrated digital networking environment, only time division multiplexers have the technical capability to handle real-time voice and video communications. For organisations that currently have a minimum of two corporate networks, one for voice and one for data, the effect of combining the two into a single integrated digital network using TDM techniques immediately brings

DIGITAL MULTIPLEXING

savings in circuit rental costs.

For organisations that use the PSTN as their main channel for voice communications to remote company locations, but that have some form of leased data links to these locations as well, the move to digital multiplexing could be just as beneficial. At the very least it would drastically reduce the use of the public telephone network by combining data and voice over the organisation's leased digital circuits. If text or image transmission play a part in corporate communications (and this would include methods such as telex and facsimile) they, too, could be incorporated into a digital network which could again reduce the reliance on public services. The other obvious advantage is the resulting cost savings, increasing the utilisation of existing leased digital circuits whilst reducing the costs associated with PSTN usage.

Until recently not many organisations had even considered the use of video conferencing as a practical solution to organising meetings and conferences. This was mainly due to the time and costs involved. Specialist studios tended to be located only in the larger cities, and the capital outlay for private studios made the facility a far from attractive commercial proposition. This viewpoint is beginning to change, given the escalating costs of moving personnel any distance, especially in terms of time and money. The other factor to be considered is the developing digital technology, which makes video conferencing a practical possibility, especially when utilising the existing corporate digital network. Studio facilities are no longer required, as compact systems the size of a large television set are now available. Full motion, full colour, two-way video conferencing gives the benefit to face-to-face communication. Simultaneous data, graphics and facsimile transmission can also be catered for.

All this can be routed through a standard 2 Mbits digital circuit, although compression techniques can reduce the bandwidth required to less than six voice channels (around 384 Kbits). Megabit multiplexers provide the interface between the megabit circuit and the video conferencing system. Other video applications include slow-scan TV. Although it does not provide all the facilities of video conferencing, it only requires a standard 64 Kbits channel, thus providing a cost-effective solution to applications such as remote security camera operation. Multiple slow-scan channels can again be routed through megabit multiplexers.

Flexible Control and Resilience

The ability to configure and alter the attributes of any software-programmable TDM or STDM multiplexer on the corporate digital network from a central control point must be a very useful feature for organisations that require a flexible corporate communications structure. Using the centralised control, the network operator can logically insert or remove remotely located multiplexers without disrupting the flow of information. This ability to reroute channels can allow field engineers to perform maintenance on remote hardware such as multiplexers while end users maintain lines of communication via alternative routes. Circuit failures can be manually or automatically bypassed in a similar manner, with little or no disruption to the end users.

If necessary the network operator can reconfigure the entire multiplexer network with just a few simple instructions to the central control software. This would be especially useful in a digital networking environment where a number of configurations may be required to gain maximum advantage from the digital circuits' bandwidth and speed. For example, a daytime configuration may be required during office hours which is optimised for a large number of fairly low-speed channels for voice and data terminal communications. At the end of the day, the network operator could load a night-time configuration which was optimised for inter-computer file transfer, enabling large batches of data to be transported quickly over a few high-speed channels. In this way the software configurable multiplexer provides the optimum solution for the available network resources, which in turn gives the organisation its most cost-effective use of corporate communications.

In addition to manual reconfiguration, automatic channel rerouteing facilities are also important in terms of network resilience to component failure. Consider the potentially large number of channels routed through any one multiplexer node. Manual rerouteing methods would be virtually ruled out due to the length of time involved in a manual operation, not to mention the numerous possible errors. Automatic rerouteing would involve one or more multiplexers actioning a predefined command file in the event of network component failure. These files would contain strings of command routines which were originally devised and stored in a command file library. Such command routines would probably take the form of defining primary and secondary alternative routes for each failed channel, together with priority levels and other relevant parameters.

DIGITAL MULTIPLEXING 51

If a fault occurred, the system would automatically reroute each channel in accordance with the command routine. This automatic operation would take place at a speed far in excess of any that could be achieved by manual operation, and with little likelihood of errors, assuming that the command routines had been properly developed and tested beforehand. The fault which initiated the rerouteing process would also be notified to the supervisory terminal, so while automatic rerouteing was taking place, the network operator could locate the fault and take corrective action.

Assuming that the alternative communication routes also contain traffic (as in most networking environments) a problem may arise. The combination of this traffic and the rerouteing traffic may exceed the traffic handling capacity of the digital circuit. In this case, some form of priority allocation may be required to enable the system to decide how the capacity that exists should best be allocated. This is usually achieved by defining priority levels for each user, to be employed in situations such as this. If rerouteing takes place, all users with high priorities are automatically given first priority in their use of resources, while the lower priority users are allocated what remains. In actual practice this could mean that low priority users on the alternative routes are disconnected or 'bumped off' the circuit in favour of higher priority users.

There are other implications regarding rerouteing that should be borne in mind by network designers. The first is the condition or time period that should initiate automatic rerouteing of channels. As rerouteing is a fairly dramatic step to take, the conditions under which it is initiated should be clearly defined. For example, a temporary loss of signal and complete line failure are two conditions which can really be measured only by elapsed time. If too short a period is stipulated for a loss of signal, rerouteing may be initiated unnecessarily. Too long a period may cause unnecessary disruption to users.

Within this equation, the time-out conditions of various attached devices such as FEPs should be considered. Such devices may require operator intervention before the users can reconnect to them. Within some vendors' multiplexer products there may be limitations on the number of channels that can be rerouted or the maximum number of intermediate nodes allowed on the alternative routes. Where neither of these problems exist, the time taken to reroute large numbers of channels should still be considered.

Hardware Redundancy

Although channel rerouteing facilities are important in terms of network resilience and disaster planning, an equally important feature of digital networks should be the hardware redundancy features within the individual network components. Hardware redundancy with regard to digital circuits themselves is basically under the auspice of the PTTs; the subscriber has little or no control over circuit provision. Only circuit duplication can provide some degree of resilience. However, the network designer *does* have influence over the amount of inbuilt hardware redundancy within components such as multiplexers. This redundancy feature is usually achieved by duplicating individual hardware components within the multiplexer (such as power supply or common logic boards) using one of two main methods.

The first method is to run duplicated components in a 'primary and secondary' fashion, with the secondary or back-up component operating in a 'hot' or 'warm' standby mode. Hot standby implies that the secondary component is immediately ready to take the full load from the primary component should it fail. This should mean that the end user notices no detectable difference if and when the switchover occurs. Warm standby, on the other hand, implies that the secondary component is not immediately ready but requires a warm-up period before it can take over from the failed primary component.

There is one significant potential problem with the primary and secondary back-up method. If and when the primary component fails, the secondary component, unused for a long period of time, may not be able to take the sudden loading, and it may fail as well. The answer to this is to preprogram the multiplexer to switch over automatically at regular intervals. This swaps the primary and secondary roles of components, as well as providing a test of the back-up function.

The other method of operating duplicate components is in a parallel mode, each with a share of the total loading. This way both are partly loaded, and if either failed, the remaining load would immediately be transferred to the remaining component. Some forms of parallel operation operate in a '60/40' split of the total load, with the multiplexer swapping the split between the two components at regular preset intervals.

The drawback with hardware redundancy is cost. Duplication of hardware components obviously duplicates cost, and in addition, extra

DIGITAL MULTIPLEXING

intelligence is required to undertake the switching feature. The amount of redundancy required therefore depends on the quality of service end users require, and reliability must be balanced against increased expenditure on communications hardware.

4 Digital Data Switching

GENERAL BACKGROUND TO DATA SWITCHING

A basic requirement of any data communications network is the theoretical need for any device within the network to be able to communicate with any other suitable device. The problems involved in achieving such a basic requirement can be summed up in one word — incompatibility. It can be found in terminal types, protocols, transmission modes and a large number of other areas. Even today the dream of interworking all communications devices within a network is very far from being a reality.

The result of this incompatibility has been the development of a number of small, specialised networks within an organisation, each with a specific application and with little or no interconnection between networks. This approach may mean that users have several terminals piled on the desk, and other consequences may include an expensive and complex local area cabling layout which increases for every new or relocated terminal; numerous point-to-point circuits between various remote locations, often running incompatible communication facilities along parallel physical links; and increasing expenditure on training for both users and network operators in the use of new communications equipment. Add to all this the complexities of the corporate voice network, and it can easily be seen why so much time, money and effort is expended on attempting to control this corporate resource.

One possible way to resolve at least part of such a problem is to find a method of removing some of the physical aspects of a multiple terminal, circuit, and cabling environment. One of the simpler methods of achieving this is to install some form of channel or circuit switching device, which together with either inbuilt or external protocol and speed conversion devices forms the basis of a standardised network. In theory the switch would allow standardisation to take place on just one or two terminal types. These in turn would be linked via a standard local cabling system and common point-to-point circuits to the central switch, which in effect would be the hub interfacing to the corporate computing, communications and IT resources.

Within certain network designs, voice traffic could also be incorporated into the overall network plan. Either the two communication methods would use the same multi-function switch, or the voice and data would be separated at central points into purpose-built voice and data switches. For an organisation with multiple locations, the same solution could be applied: each location would be equipped with one or more switching devices. The next step for such an organisation would be to link remotely-sited switches together to form a wide area corporate switched network.

The most familiar example of circuit switching over a wide area can be found in the public switched telephone network (PSTN) provided by the PTTs. Its switching abilities form the basis for the service we know today. Although this is traditionally an analogue-based system primarily for voice traffic, data can also be transferred across the network. The advent of the public Integrated Services Digital Network (ISDN) extends this switching concept over many different communications services, changing the network emphasis from voice to a combination of switched voice and data communications.

This same design concept is a requirement of many corporate networks, although in most cases on a far smaller scale. The economies achieved by reducing the number of leased circuits between sites, reducing cabling requirements within sites, and generally improving utilisation of the remaining physical links, are just one of the factors that favour the installation of switching devices. Indeed, these cost-saving factors alone may well justify the need for switches even when weighed against the high capital investment required during the initial switch installation phase. As with the PTTs, organisations have seen that the use of digital technology in both switching and transmission techniques is far more cost-justifiable than any analogue equivalents, especially when considered over a reasonable timescale.

If the switch is the answer to part of the network incompatibility

DIGITAL DATA SWITCHING

problem, the next hurdle is what type of switch to install. The switching concept has evolved and is still evolving from a number of different backgrounds. The main digital switch types are briefly outlined below before they are reviewed in more detail later in the chaper. They are:

The megabit circuit switch: Originally based on the requirements of the PTTs, the automatic circuit cross-connection equipment is capable of handling large numbers of megabit circuits, providing individual timeslot channel cross-connection at the 64 Kbits level. The manufacturers subsequently developed smaller versions to provide megabit switching for large corporate networks.

The kilobit matrix switch: From the manufacturers of the simple hardwired 'A to B' circuit switch has come the electronic, software-based kilobit matrix switch. This automatically performs the same functions as the manual hardwired circuit switch or patch panel, but in a fraction of the time.

The data PBX: The traditional manufacturers of communications equipment such as modems and multiplexers saw the need for a data switch which would integrate into their mainstream product range. Their approach took two routes: to provide switching capability within their more advanced multiplexers; and to provide a stand-alone network product generally termed the data PBX which would provide switching and other facilities, as well as integrating into their existing proprietary equipment user base.

Front End Processors (FEPs): The computer manufacturers originally developed FEPs and networks concentrators to handle their proprietary networking requirements. One function resulting from this was the ability to provide data circuit switching. This was further enhanced to include gateways and support for other proprietary networking architectures, and in many cases such devices could now be classed as switches in their own right.

PABXs: The traditional voice communication PABX manufacturers, with a long experience in switching technology, began developing exchanges which would handle both voice and data over one integrated network. This has resulted in some cost-effective local area network solutions and a growing presence in the digital wide-area data communications field.

All these switch types have advantages and disadvantages depending on the network design and applications required. In general, each type has a specialised use but has areas of overlap with the others. To aid in the choice of switch, we will look at some of the technical aspects of each followed by some of the pros and cons involved in its use in a corporate network.

TECHNICAL CONCEPTS BEHIND DIGITAL SWITCHING
The Megabit Circuit Switch Technology

The need to provide for 2 Mbits circuit switching in today's corporate network is not at first sight a requirement of many organisations. Indeed, such a capability is currently only applied within the PTT networks and the very largest private networks. However, as the use of 2 Mbits circuits becomes more widespread, the demand for this type of switching ability may well percolate down to other organisations as their communication needs become more complex.

The original specification for megabit circuit switching was drawn up to meet the requirements of the PTTs. They saw features such as rapid circuit provision, circuit routeing flexibility, remote configuration and control, and several levels of component redundancy as essential aspects for a megabit circuit switch within their internal network. From this specification came the Automatic Cross-connection Equipment (ACE).

The ACE allows the PTT the respond quickly to customer requests for the installation of digital circuits. This in turn is beneficial to the corporate network designer, who can design and alter the private digital network in the knowledge that the PTT circuit provision is both readily available and flexible.

To look at the ACE switching system in more detail, let us examine its use in a PTT network. At the local PTT exchange, individual 64 Kbits circuits coming from a number of subscriber premises are multiplexed together onto a 2 Mbits bearer circuit by a sophisticated time division multiplexer known as a *primary multiplex*. This 2 Mbits bearer circuit (supporting up to thirty subscriber timeslot channels, each of 64 Kbits) is directed to a cross-connection site along with similar circuits from the local exchange area and 2 Mbits circuits routed directly through from subscriber premises. Assuming the site is equipped with ACEs, the circuits are terminated at these devices and presented to the ACE in a standard CCITT G.703 co-directional interface form, using HDB3 coding.

Inside the ACE, the 2 Mbits circuits are demultiplexed back down to the individual 64 Kbits channel level. Each channel is then routed through the device according to a predefined path to the appropriate output channel, where it is again time-division multiplexed with other similar 64 Kbits channels onto a 2 Mbits bearer circuit for the onward journey across the network.

As the ACE is a software configurable device, it can be locally controlled by attaching an asynchronous terminal directly into a supervisory port. More importantly, however, it can be remotely controlled and coordinated along with other ACEs within the network from a central control centre using a supervisory system known as an *RCE* (Remote Control Equipment). The RCE performs the functions associated with control and management of the network. The device is based around a processor ranging in size from a single microcomputer up to a large minicomputer, and the computer's size is indirectly related to the size of the network it has to control. The RCE supports a large database which contains detailed information regarding circuit routeing and provisioning, as well as such details as the status of all network equipment and lines.

The RCE can support a number of supervisory terminals from which network operators can access the database in order to set up, modify or discontinue a circuit on any of the attached ACEs, even when it is necessary to coordinate a number of ACEs to achieve this (see Figure 4.1). Similarly, the network operator can use the supervisory terminal to undertake maintenance tasks, testing circuits by cross-connecting sections of a circuit into remotely controlled test equipment. Again this utilises the database as a reference and control facility. This ability reduces the time taken to localise and correct circuit faults. The RCE also undertakes the circuit performance and alarm monitoring functions for the whole network. This includes monitoring the network's kilobit and megabit circuits, as well as the ACEs themselves.

The supervisory network link between the RCE and the ACEs is normally achieved by providing a pathway physically separate from that of the primary data circuits. In the case of British Telecom, this means that the PSS network provides the supervisory links between RCEs and the ACEs; the ACEs themselves provide cross-connection facilities for services such as KiloStream and MegaStream. As an alternative to using packet switching as the supervisory link, the existing analogue PSTN, or

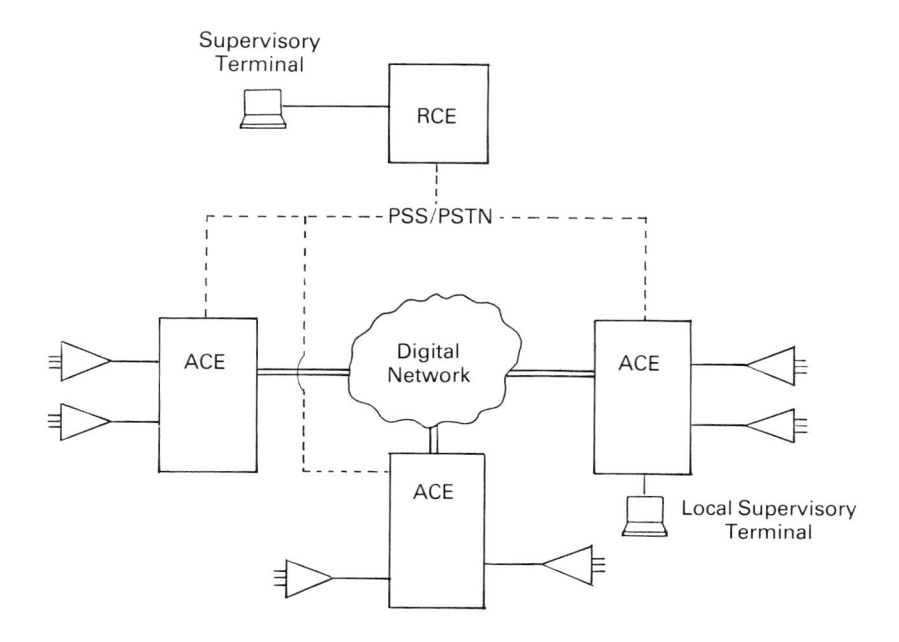

Figure 4.1 The RCE Coordinates ACEs within a Digital Network

channels within the primary data circuits themselves, can be used for supervisory purposes.

The ACE incorporates a non-volatile memory which is used to hold and maintain a current record of the ACE's cross-connection map. The map is a logical picture of where, when and how circuits are cross-connected within the device. Any request to alter circuit configurations within the ACE also updates the ACE's map. In the event of ACE failure, the map is retained within the non-volatile memory so that when the ACE is brought back into operation it can automatically configure itself according to its map. The map is also used by the RCE to cross-check its own overall map of all circuit routeings and ACE configurations. This provides additional security for an ACE: it has access to a master copy of its own configuration if required.

If the supervisory link between the ACE and the RCE failed, or if the RCE itself failed, the ACE would continue to operate using its current map until the supervisory link was restored. This means that existing

DIGITAL DATA SWITCHING   61

users would be unaffected by any failure in the control and monitoring system.

The Kilobit Matrix Switch Technology

The kilobit matrix switch is a modern descendant of the earlier manual patching panel. The manual patching panel was originally introduced in the early 1960s to provide a reliable way of switching analogue data circuits. This was achieved by physically unplugging, plugging and cross-connecting patching cables between the various circuit interfaces on the panel. The drawback of manual patching was the fact that it was just that — manual. For large and growing networks it was a slow and cumbersome method of circuit switching, especially if it was required on a fairly regular basis. The development of an automatic, solid state electronic matrix switch was therefore necessary to the overall development of corporate networks. It removed the need to physically alter cabling, moving instead to circuit switching based on intelligent central control.

As digital communications technology was introduced, so the manufacturers of matrix switches evolved their products to handle the new kilobit transmission circuits. In technical design and practical operational use, the kilobit matrix switch is very similar to the megabit circuit switch, except that the kilobit circuit switch handles kilobit circuit interfaces directly. This provides switching facilities for 64 Kbits type circuits, although speeds of up to 256 Kbits can now be supported by such switches. In a networking environment, the kilobit matrix switch will support from a few dozen to a few thousand kilobit circuits, each directly connected from the NTU to the DCE side of the switch. The DTE side of the switch is connected to communications devices such as FEPs, digital multiplexers and other similar devices (see Figure 4.2). The result is the ability to switch any circuit between any device, giving the network operator flexibility in rerouteing communications links in the event of circuit failure, bypassing faulty FEPs, etc.

The supervisory terminal allows the usual control and monitoring functions to take place, as well as test and diagnostics routines. These may include tests such as monitoring throughput on established connections, test routines for newly installed ports or circuits, and internal and external loopback tests. The supervisory software may also have an option to interface with a proprietary network management system, thereby enabling numerous switches to be controlled from a single point.

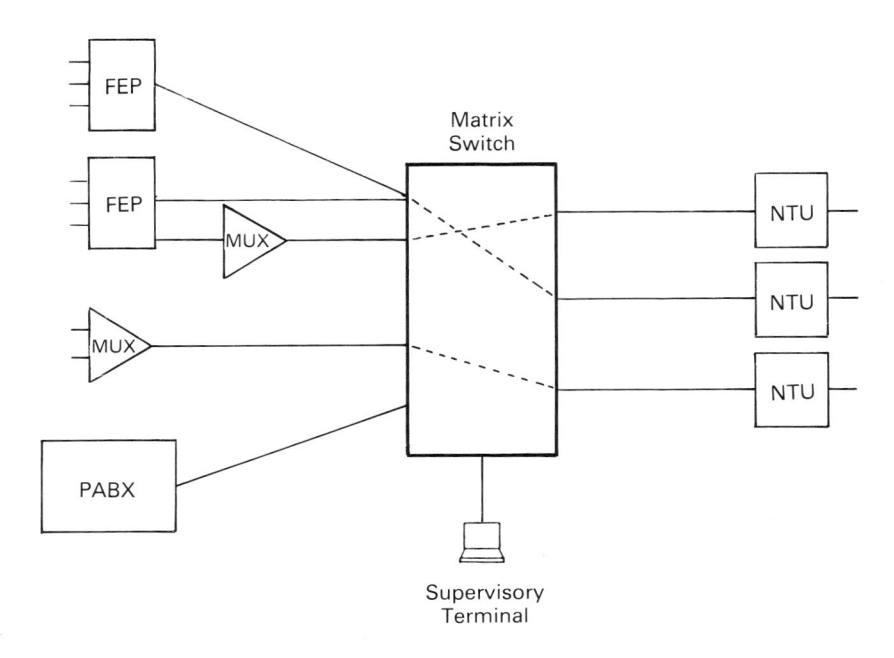

Figure 4.2 The Kilobit Matrix Switch

The kilobit matrix switch is to all intents and purposes a purely DCE-to-DTE circuit cross-connection device, an up-market version of the simple patch panel. As such, the switch is totally transparent to all speeds, codes, protocols and control signals (including network clocking signals) used within a corporate digital network.

The Data PBX Technology

The data PBX (Private Branch Exchange) has been around in a variety of guises since the early 1970s. From that time it has been marketed under various names including port selector, private automatic computer exchange, intelligent switch and data switch, but most recently as the data PBX. This latter name reflects its relationship to a private automatic branch exchange (PABX). Like the data PBX, the PABX can switch connections among a variety of attached peripherals. The major difference between the two devices is that the data PBX handles only data

DIGITAL DATA SWITCHING

traffic, while the more traditional PABX handles voice alone, or voice with support for data.

The data PBX is therefore best described as a digital data exchange where end users can contend for, connect to, and switch between all the resources attached to the switch, such as the computer ports or digital channels linking to other remote sites. The data PBX may well support as few as two dozen or as many as 4000 attached DTE and DCE devices. Within this physically attached total, a large percentage could be functioning with simultaneous logical connections. The exact number of parallel operations depends on the type of application and the combinations of data rates required. The data PBX may also be connected to several other similar proprietary switches, located locally or linked to remote locations via high-speed digital circuits. This will increase the total network handling capacity to several thousand logical connections, both locally and over a wide geographical area (see Figure 4.3).

The switch is normally software controlled, and its configuration and management are undertaken from an attached supervisory terminal or microcomputer. Additional options such as line drivers, port interfaces etc, are normally added simply by inserting plug-in cards into the framework of the device. No physical switch strapping is normally required. Through the addition of internal or external multiplexers and a digital circuit, the data PBX can communicate with a variety of remotely located equipment. Support of modems extends this remote support facility to analogue communications. The switch could provide access to X.25 packet-switched networks, and may well include protocol conversion to enable gateways to be established to other synchronous or asynchronous networking environments. Emulation of mainframe and minicomputer protocols is another possible option, providing compatibility for terminals such as the IBM 3270 or ICL 7500 series. This option may well function with a speed conversion facility, again within the data PBX.

FEPs and Concentrators

Most major computer manufacturers today provide some form of data switching capability within their proprietary network architectures. This usually takes the form of a purpose-built front end processor (FEP) device capable of supporting and switching connections between attached

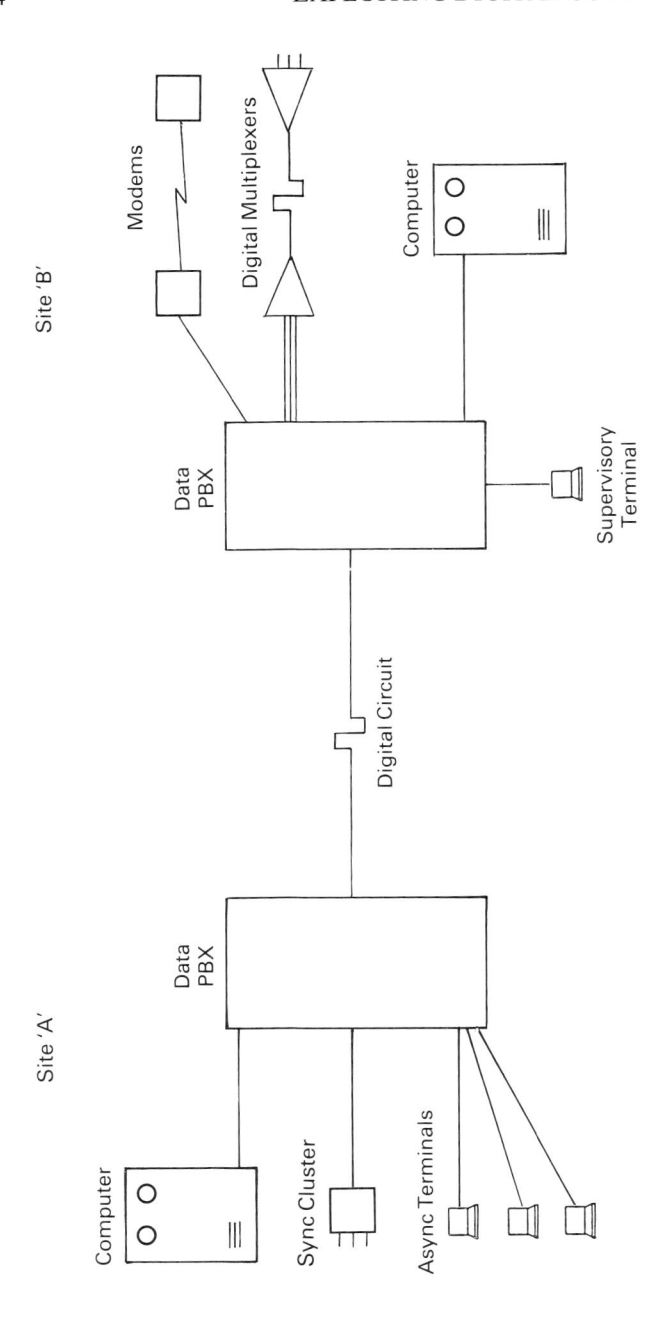

Figure 4.3 The Data PBXs Can Be Interlinked via Digital Circuits

DIGITAL DATA SWITCHING

computer processors and DTE/DCE equipment such as modems, cluster controllers or other FEPs. The IBM 3725 Communications Controller, for example, provides support for a number of attached links, with high-speed attachment of up to 256 Kbits giving a high-speed digital synchronous channel for SDLC or BSC protocols. However, FEPs in general are not particularly flexible in their design; their primary functions are to handle data communications networks, and their switching capabilities are confined to the logical routeing of data channels through the device.

The *network concentrator*, on the other hand, is a more flexible multi-purpose device. This general term is usually used to describe a minicomputer operating as a communications gateway or node on a network. Running network software, the computer may be dedicated to providing handling and switching facilities to attached terminals and other devices. Alternatively it may be multi-tasking, with the network concentrator function as just one of its tasks. Either way such a computer is normally a component or node in a larger network, interlinked to other similar minis or central mainframe computers. It provides local and remote terminals with a route through the network to the required resource, and also provides 'concentration' of low speed channels onto high-speed links. Most manufacturers provide such devices, which include the IBM Series 1, Honeywell's DPS6 and the Unisys DCP minis with appropriate networking software.

The main limitation of network concentrators is that they are modified minicomputers running network emulation software: this could make response and throughput slow in comparison with other types of switch. As a general rule, the greater the number of terminals attached to the networking function, the greater the response times.

The PABX

The traditional voice switch, the *Private Automatic Branch Exchange* (PABX), has evolved over the last few years into a computerised, digital device technically capable of providing switching facilities not only for corporate voice traffic but for data communications traffic as well. This facility can be provided on either a local or wide area basis, with digital communication links as an option in all cases.

The use of the PABX for data communications does have its advantages. These include reduced cabling requirements, increased utilisation of an existing asset such as the PABX, and some of the features described earlier under data PBXs. The whole area of digital PABXs is covered in more detail in Chapter 9, The Digital PABX, so at this point, we will concentrate on data transmission via a PABX, and in particular the main problem areas associated with such techniques: the transmission speed and switching capacity of a PABX.

The transmission speed of a PABX relates to the data transmission rate obtainable from a PABX extension to the PABX itself. On the older analogue and early digital PABXs the maximum rate was usually around 9.6 Kbits (although 19.2 Kbits can be obtained in certain cases). The latest digital designs allow for transmission speeds of 64 Kbits across a fully digital extension circuit. Links from the PABX to the corporate wide area network (and the outside world in general) can also be provided at 64 Kbits. This is achieved via individual kilobit circuits, or more commonly by utilising 2 Mbits circuits to provide multiple 64 Kbits channels.

The switching capability of a PABX relates to the way in which a number of data transmission extensions together with the voice traffic would be handled and routed within the PABX.

Traditionally the PABX has been configured on a reasonable assumption: it is unlikely that everyone will want to call everyone else at the same time. A degree of circuit concentration is therefore built in. Groups of extensions are terminated at an internal switching element, and only a few links are carried from this element to other internal switching elements. Thus the way in which the switching elements are connected and arranged affects the number of calls that can be in progress at any one time. This limitation would apply regardless of whether the calls were using speech or data. With this type of design, data calls could adversely affect the PABX because they have different call parameters to voice calls (ie data calls are generally of longer duration, and call distribution tends to be confined to a group of extensions, certain times of day etc). A PABX with concentration is often referred to as a 'blocking' PABX. This term is used because although the extension may be free, the caller may not get through as the links between one or more internal switching elements are busy (ie blocked). This blocking effect can be expressed in terms of the traffic capacity of the PABX, where the unit of measurement is an *Erlang*. (An Erlang is a measure of the density of telephone traffic. Generally, one Erlang is defined as one circuit in continuous use for one hour.) A blocking switch will therefore have a

DIGITAL DATA SWITCHING

maximum traffic capacity of 'n' Erlangs, ie 'n' simultaneous calls per hour.

The ideal state on a PABX is where each extension has a link between each switching element at any given time. In such a case, the PABX could be described as 'non-blocking'. Digital *Stored Program Control* (SPC) PABXs are now available. They are described by their manufacturers as non-blocking, and are suitable for both voice and data traffic. Such switches can nevertheless be affected by data calls of a different type, those of very short duration. This is because the PABX's central processor, which manages the set-up and control of every call, can be overloaded by a rapid succession of short data calls (eg those of numerous PCs using the PABX to access records off a central file server). Data calls are on the whole less efficient over circuit switching networks.

USER ISSUES REGARDING DIGITAL SWITCHING
ACE in a Corporate Network

As I originally mentioned, the ACE was developed primarily for PTT use, and has the capability to support up to 128 2-Mbits circuits. To meet the needs of corporate networking, a mini-ACE was introduced to supply similar functionality on a smaller scale, supporting up to 32 2-Mbits circuits. Routeing for individual 64 Kbits channels can be provided on several bases, including unidirectional point-to-point circuits, both-way point-to-point circuits, unidirectional multi-point circuits and 'n $\times$ 64 Kbits' channels (eg megabit channels). All circuits and channels are totally transparent to user protocols and formats.

The ACE's main function is as a central backbone switching device controlled and managed by the network operator. To apply such a device to a corporate network would probably mean a requirement for extensive use of wideband applications such as broadcast transmission, video conferencing, and high-speed data transmission between numerous computer centres. It might also be used as a switching function in a large corporate voice network interlinking PABXs. The central control for ACE channel provision may be viewed as a disadvantage by some corporate network designers. This is because the end user would have to contact the central network operator each time a new channel route was required. An obvious development for the future would therefore be a facility enabling end users to set up a route through the network locally, similar to the facilities offered by data PBXs.

The ACE is built to a very high standard of reliability and performance demanded by the PTTs. Hardware is duplicated or triplicated to ensure system resilience. This means that the mean time between failure of a large ACE is in the order of at least 100 years (a failure being defined as a loss of service to 50 percent or more of the circuits). As regards performance, time delay across the ACE is in the order of one frame, about 390 nanoseconds. Even at this performance level PTTs tend to limit the number of ACEs through which a channel passes to around six. High cost is an inevitable result of such high standards and performance, so a version cut down to normal corporate networking requirements is being produced to provide the same functionality at a competitive price.

The Matrix Solution

The kilobit matrix switch has many similarities to the megabit ACE in its design, operation and application requirements, as well as in circuit and channel transparency for user protocols and formats. The kilobit matrix switch is designed to be as resilient as possible, which is important given its strategic position in most corporate networks. It is likely to be in a critical central role, interfacing to many other communications devices such as PABXs, FEPs, multiplexers etc, and probably numerous digital circuits as well. Its operations (including circuit switching, monitoring and managment functions) are all centrally controlled by the network operator. This approach makes the device a centralised network component rather than a facility accessed and manipulated by the end user.

Its strategic position and application within a corporate network tend to suggest it is a device designed for and marketed to the larger organisation (ie an organisation whose yearly purchasing requirement would place it in the top 10 per cent of the overall UK customer base for communications products). To cater for smaller organisations, however, kilobit switches as small as 4 × 4 switching interfaces have been introduced. The manufacturers of kilobit switches are also moving into the megabit market, providing megabit matrix switches which will compete with the smaller ACE devices.

The Data PBX Applications

The data PBX has an appeal to two major interest groups. The first group are those organisations that have a large number of terminals which must

DIGITAL DATA SWITCHING 69

communicate with local or remotely sited computers, or with other resources such as word processors, printers and microcomputers. It should be noted here that although a physical connection can be established, interworking may be limited due to protocols or other limitations. This is why most data PBXs have the option to provide some form of protocol, speed or other conversion facility to assist in device interconnection.

The second group with an interest in data PBXs are those that already have a large investment in voice PABX technology, or that view integrated data and voice PABXs with some concern. For these organisations voice and data communication is handled completely separately, or at best with some limited interconnection to allow 'data-over-voice' modems or multiplexers to be used on voice circuits, while maintaining independent operation through the use of separate data PBXs and voice PABXs.

The appeal of the data PBX is that it was and is designed purely as a data switch for data applications only. It does not try to compete with the complexities of the integrated voice and data solutions, but in its own field it does have a specific function. The data PBX has been compared in functionality with the advanced digital multiplexer, or more commonly with the Local Area Network (LAN). When data PBXs and advanced digital multiplexers are compared, it is found that both occupy a slightly different niche in network design. Whereas some digital multiplexers provide switching for a variety of DTE and DCE devices, their primary application is in accessing remote resources. A data PBX on the other hand is possibly less concerned with remote connection, but is ideal for handling a large amount of internal throughput, thereby avoiding transmission delays when large numbers of devices are connected. For local data switching with an option for remote access the data PBX fulfils the requirement more efficiently and often less expensively.

The data PBX also offers a low-cost alternative to local area networks. Although the LAN has several features not yet available to the data PBX (such as megabit transmission speeds, the ability to perform file transfer functions, and the means to fully distribute computer resources) the data PBX does offer some interesting advantages. Many organisations do not require high-speed transmission. The data PBX's non-blocking architecture ensures a slower but steady throughput performance as long as its capacity is not exceeded. Some LAN systems suffer from performance degradation under heavy loading conditions as a matter of course. Smaller organisations with limited technical personnel have found the relative simplicity of a compact data PBX an advantage over LAN systems, which require a great deal of pre-installation planning and ongoing maintenance. Another advantage of the data PBX is its comparative cost. Viewed on a purely price-per-port costing, the data PBX can generally offer a port connection at a fraction of the cost of a LAN equivalent.

Some organisations view the data PBX and LAN technologies as compatible rather than competing. They have high-speed digital LANs interconnecting various computing resources, but at the same time have data PBXs supporting the local interactive asynchronous terminal environment. Both technologies have gateways and bridges to provide interconnection. This approach to interconnection has been taken up by some manufacturers, who see the two technologies as part of a wider networking environment and provide products to meet this requirement.

The FEP Solution

If an organisation's data processing department is of any size, the corporate data network will almost certainly include one or more FEPs connected to central host mainframes and/or a distributed network of minicomputers operating some form of network concentration procedure. The question in such a case is whether the organisation should expand and develop such equipment to cope with high-speed digital communications and switching, or whether alternative switching options should be considered.

The answer will be unique to each organisation. An upgrade to existing FEPs and network concentrators will in most cases be cheaper than installing other types of switch. However, this approach implies a proprietary solution with a possible loss of flexibility in the choice of future suppliers and possible restriction in the development of future applications. The extra cost of installing communications-oriented switches could therefore be offset against the flexibility, integration and 'futureproof' arguments. It is a solution that could adapt to future requirements without making the existing switching equipment obsolete, and which allows data communications to be combined with the corporate voice, text and image needs.

The Integrated PABX Solution

For some years now, the manufacturers of digital Private Branch Automatic Exchanges (PABX) have been promoting the device as *the* solution to an organisation's networking needs. 'Integration' is the key word. PABX offers the user the potential to integrate voice and data communication, as well as text and image if required. This is a big step away from the still traditional perception of the PABX as a purely voice-oriented switch.

The main advantage of using the PABX in an integrated fashion would be the cost savings involved. One type of switch could be used for all communication methods, thereby avoiding hardware duplication. Just one set of cabling could be used, thus providing an environment that allowed any communications device to be transported to any suitable extension socket, cutting cabling costs dramatically. One of the longest-running debates within the telecommunications world is whether the digital PABX is capable of handling data communications. Many PABXs were originally conceived as devices primarily dedicated to handling and switching voice communications. Queueing and probability theories used by the designers assumed that voice calls would be made randomly, and that an average call would last around three minutes.

Data calls are in many ways a complete reversal of this theory. Many data calls established through a PABX could last for at least an hour, and possibly for the entire working day. Because of the requirements of computer applications, the number of data calls during the working day tends to create peaks at certain time periods rather than being randomly distributed. Similarly, throughout a week or month peaks may occur on certain days. A third point to note is that computer communications are tending to require more bandwidth throughout the PABX; 64 Kbits channels are now a common requirement for many processor-to-processor applications.

The result of these longer calls on the performance of the older types of PABX can be dramatic. With relatively few extensions engaged in data communications, the system can in effect become overloaded. This means that the data calls effectively 'block out' calls to other extensions, so that callers receive an 'engaged' tone even though the extension they are calling is not in fact in use. It has been estimated that most older designs of PABX begin to suffer from overloading when more than about

5 percent of their extensions are in use for data calls. As well as the problems caused by longer data calls, another set of problems may occur if the opposite situation arises. The growing use of office automation equipment networked through a digital PABX has resulted in an increase in very frequent data calls of only a short duration. Take, for example, an organisation with a number of personal computers which are being used together to handle related tasks. Any of the computers may well need to refer to any other at any time for just a small quantity of information held within a particular database. These extremely short data calls cause two main problems. First, being geared to the needs of traditional voice calls, older types of PABX tend to take several seconds to switch through a call. This is obviously very inefficient when the actual duration of the call will only be in the order of a few seconds. Secondly, the control systems of a PABX are normally designed to set up only a finite number of calls every second. If the number of computer systems requesting the set-up of data calls is very frequent, conventional voice callers may very well be 'locked out' of the telephone system, unable to obtain even a dialling tone.

To counter the 'blocking' effect, a new design of digital PABX was introduced which had a 'non-blocking' internal organisation. This non-blocking approach basically utilises statistical time division technology so that time slots within the device are utilised to maximum efficiency. A non-blocking switch is one that can provide its maximum specified number of throughput channels, each potentially operating at its maximum defined data rate, without any degradation. Other switches without this feature are considered blocking (ie the number of throughput channels is reduced as the data rate per channel is increased).

The distributed approach to PABX design can also (in some cases) enable the effect of data calls to be spread across the network. This is achieved by careful network design, ensuring that the local node of a distributed PABX has the capacity to handle the local data communication needs. Local workstations, computers and PCs are thus routed on a local basis, leaving the rest of the distributed system to handle the voice requirements and the remaining data communication needs. The distributed PABX nodes are then interlinked by megabit circuits to provide a complete system.

In general, as the requirements of data communications grow, the PABX manufacturers attempt to keep pace by introducing faster, more powerful devices to meet customer needs. However, the problem in many cases is that the PABX is at least one step behind the requirement of the organisation concerned. The current view of many organisations is that although the PABX does have a place in data communications, it is not as yet the complete answer.

5 Digital Communications Applications

As we have seen, digital multiplexers and switches are based on technology which was originally developed for the requirements of analogue communications. However, there are a number of new devices now on the market which were specifically developed for digital communications technology. These make use of the resulting benefits to provide a number of new services and facilities to the corporate network designer. The following chapter reviews some of these products, looking at desktop video communications, SNA communications operating at megabit speeds and the attachment of computer peripherals over megabit circuits.

DESKTOP VIDEO COMMUNICATIONS

Until recently video conferencing was an expensive communications method which required purpose-built studios, complex equipment and very wide bandwidth communications links. Such systems were therefore confined to large multi-national organisations, or organisations that supplied this communication system as a commercial service to others. Today, all the technology and expense has been reduced to a desktop video communications device costing in the region of £10,000. This has put video conferencing technology in the reach of a number of organisations that could benefit from its application within their business.

One version of desktop video communications bases its user/network interface around what at first sight looks like a normal personal computer. Indeed, this video workstation can be used as an IBM compatible PC in its own right, with all the usual PC facilities such as local processing capability, colour graphics display and interfaces for local area network data communications. In addition there are video communications facilities. The PC monitor doubles as a high-resolution colour video screen, and a colour camera, viewfinder and microphone/speaker are built into the workstation enclosure.

The video communications system is controlled by the user, who can select the operating mode of the workstation, and then, using a normal telephone-like keypad, make and establish full-motion, two-way video calls to other suitably equipped users on the network. Multi-way conferences can also be set up, with the workstations automatically switching to the current speaker's voice. Additional local video inputs and outputs on the workstation allow the attachment of an auxiliary video camera or video recorder, or output to a large screen. A telephone handset can also be attached to give privacy to video calls, although it does not give connection to a PABX or the PSTN.

The video communications system works by linking the video workstation to a locally sited cluster controller. The physical link can be up to 1000 feet using RG-59/U coaxial cable or up to 1750 feet when using RG-11/U cable. Both cabling types allow the video communications signal to be combined with any LAN data traffic for the associated PC. The cluster controller itself is a small non-blocking matrix switch capable of supporting up to eight locally sited video workstations. Its switching function is controlled by any of the attached workstations. To form larger video networks, the cluster controller can be interfaced to a broadband LAN, with up to 32 cluster controllers being linked together over a building or site. Ports on an IBM 37X5 Front End Processor may also be connected to the LAN to enable 3270 communications to take place between IBM mainframe hosts and the video workstation's associated PC. The PC can also connect to baseband LANs if required.

Each single-way video signal takes on average 6 MHz (megahertz) of bandwidth over the analogue-based broadband network. This can be compared with analogue voice transmission, which can often be carried out using only a 3 KHz (kilohertz) channel. To establish a two-way video and audio channel with the associated system signalling would therefore require just over 12 MHz of bandwidth.

For wide area video communications, a cluster controller can be interfaced through a device known as a *codec* to either megabit or kilobit digital circuits. These circuits may be integral to or independent of the corporate network structure (Figure 5.1). The codec performs two important tasks. First it converts the local analogue signal to a digital

DIGITAL COMMUNICATIONS APPLICATIONS

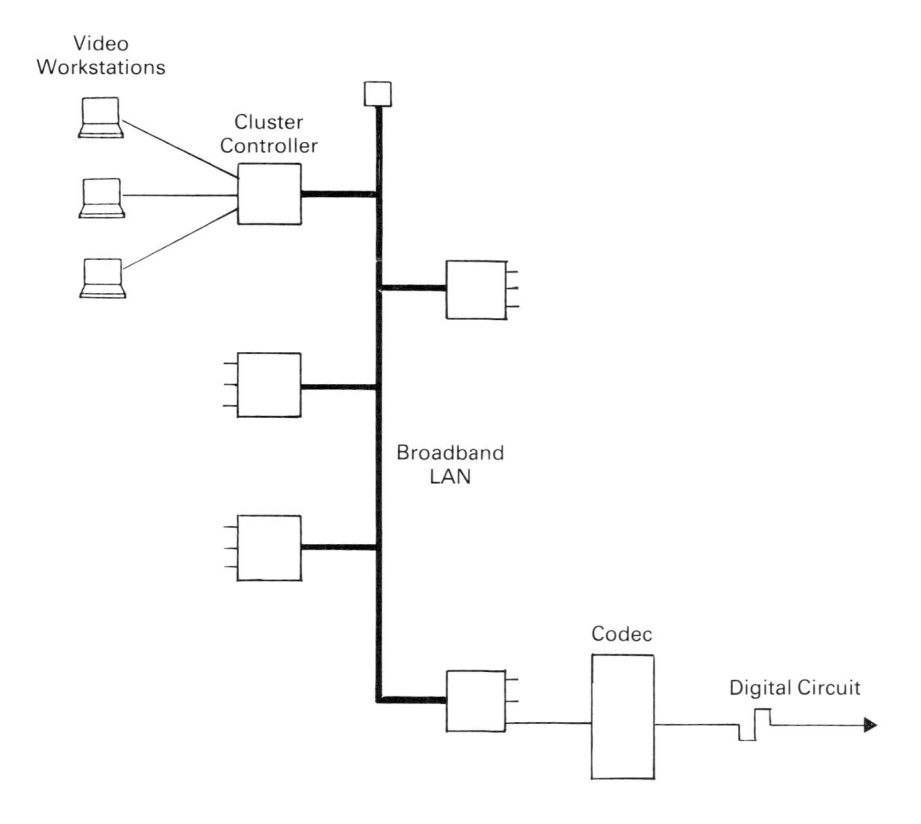

Figure 5.1 Desktop Video Communications System with Links to a Wide Area Digital Network

format for the wide area circuit. However, a simple conversion of just a single 6 MHz analogue video channel would take in the order of 90 Mbits of bandwidth on its digital counterpart, so the codec's second task is to compress the digital signal to a fraction of its original size. The resulting bandwidth will be around 64 Kbits to 2 Mbits, depending on the picture quality required. The faster the speed, the better the quality of picture, although the final perceived picture quality is (as usual) subjective. This compaction technique for digital video signals is generally known as *compressed video*.

In the system described above, the speed normally recommended is 384

Kbits, giving a good quality picture at a reasonable bandwidth usage. In such cases, a number of video channels could be routed down one 2 Mbits circuit, or alternatively, with the use of multiplexers, video channels could be integrated with other corporate communications such as voice, data, fax etc. The system could also be connected to a switched digital service from a PTT. One of its first applications in this area was over the AT&T Accunet Switched 56 service in the US, where the video communications system operated over the 56 Kbits switched public digital network.

Given that the technology of desktop video communications exists today, what are its applications in a business environment? Its primary benefit is to reduce the need for face-to-face meetings, saving the time and cost of bringing people together. Live image, transmission of diagrams or written ideas using an auxiliary camera, and simple data file transfer can all be combined to allow meetings to take place from each person's office. With this system, executive meetings, project reviews, seminars and presentations can all be held with the minimum of effort for the participants. These benefits have to be set against the cost. Cluster controllers, cabling, codecs and circuit leasing costs all have to be considered in addition to the video workstations. However, at least some of the networking components may already be in place with a corporate network. Video communications could therefore become another facility for the corporate user.

MEGABIT SNA COMMUNICATIONS

Proprietary communications processors are available which allow IBM compatible mainframes and remotely sited 3270-type terminals to be connected via a 2 Mbits digital circuit. The effect is to make remote terminal response times equal to those of locally sited terminals that are directly connected to the host machine. The system works by channel-attaching the host computer to the locally sited megabit communications processor, which effectively replaces or operates in parallel with the existing front end processor. Several configurations of the system are then available. One version allows up to four 2 Mbits circuits to be connected to the megabit processor, these circuits being terminated at the far end by proprietary 3270-type terminal controllers. Up to 32 terminals can then be attached to each controller, giving up to 128 remote terminals operating over four high-speed circuits (see Figure 5.2).

An alternative configuration is to have a single 2 Mbits circuit

DIGITAL COMMUNICATIONS APPLICATIONS

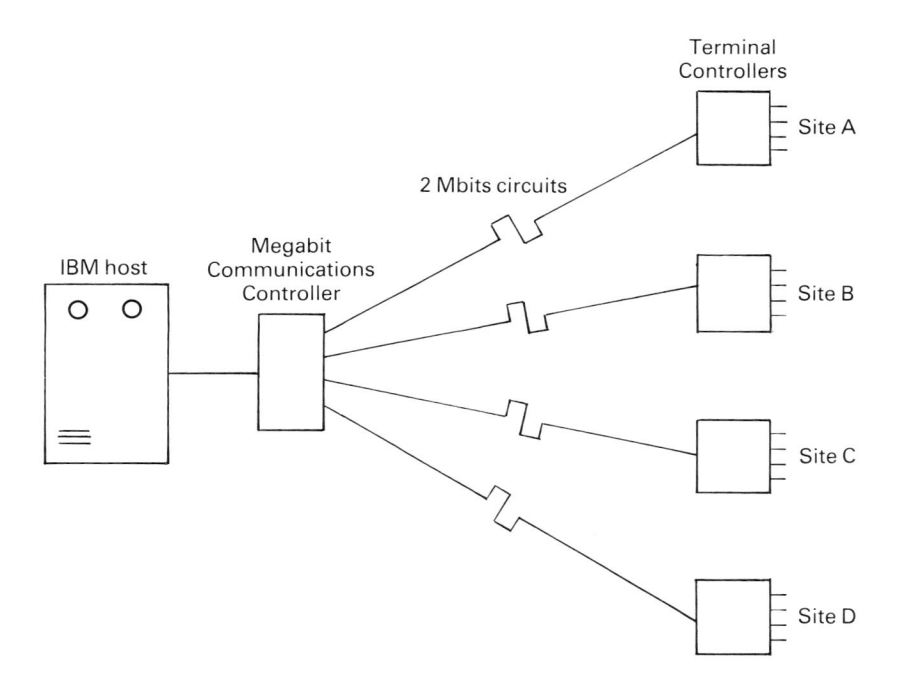

Figure 5.2   Remote Terminal Support over Megabit Circuits

supporting a remote circuit concentrator, which in turn supports a number of terminal controllers. This could allow a 2 Mbits circuit to replace a number of existing analogue or kilobit links, while at the same time increasing the response time of all the remotely connected terminals at that site (see Figure 5.3). Although such systems are not cheap, they do provide an alternative to a situation where either relatively slow response times would exist for remotely sited terminals, or where additional computers would have to be installed on remote sites to provide the necessary local processing performance.

CHANNEL ATTACHMENT OVER MEGABIT CIRCUITS

Many large organisations with multiple computer sites often have a requirement to access information or other resources which are spread over the various computing locations. The traditional solution to this access problem has been to interlink the sites with a data networking

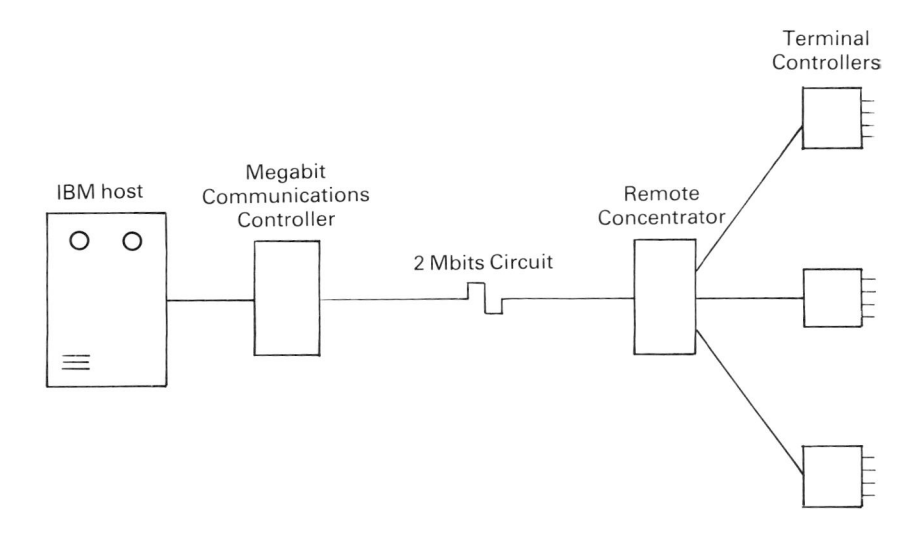

Figure 5.3 Remote Terminal Support over a Single Megabit Circuit

architecture that allows real-time user access or permits bulk file transfer between the remotely located hosts. An alternative to this traditional approach is the use of a high-speed communications device that allows peripheral equipment, which is normally channel-attached to a local host computer, to be sited at a remote location but still provide channel performance comparable with similar locally sited peripherals.

The system works by linking the locally sited host directly to the high-speed communications device, using the normal peripheral input/output channels. At the remote site, an identical communications device is linked to the required peripherals. No software or firmware changes are required to either the host computer or the peripherals, since each appears as a normal, local channel-attached link to the other. The interface between the two remote sites is provided by a 2 Mbits digital circuit, terminated at each end by the high-speed communications devices. In operational use, the system allows the local host computer to have high-speed access to remote peripherals such as laser printers, line printers, CAD/CAM devices and tape drives, the access being provided without the use of any remotely sited host or FEP. If remote peripheral sharing is required between a suitable remote host and the local host

DIGITAL COMMUNICATIONS APPLICATIONS

computer, this can be achieved by the addition of a channel switch at the remote site.

The system allows maximum usage to be made of the available peripheral resources, at the same time reducing the need to install back-up equipment at each site. Instead standby facilities are provided through remotely sited peripherals. New procedures could be established whereby data held on mass storage devices could be copied and secured to remotely sited peripherals, without the need to supply the remote site with a processor or even personnel. For some time computer manufacturers have provided the ability to link local processors on a close-coupled, channel-to-channel basis, but 'local' has always meant that the machines are located within a few hundred feet of each other. This system offers a way of providing channel-to-channel attachment for two remotely sited hosts.

The main constraint on such a system is the effective speed of the megabit link. A 2 Mbits digital circuit is full duplex and therefore capable of transferring a 2 Mbits data stream in both directions simultaneously, giving a theoretical maximum aggregate rate of 4 Mbits per second. In comparison a local channel of a host machine is capable of around 6 Mbits per second, although usually only in half duplex mode. It was therefore not until the arrival of the 8 Mbits digital circuit that such systems could operate to their maximum potential.

One of the earliest applications for remote host interconnection was undertaken by a UK financial organisation. Two large remotely sited computers together supported an equally large real-time network. Initially each site updated its own database in real time, but also held a recent copy of the other's database. This was off-line updated by copies of the latest transaction update files transmitted over the 8 Mbits link, allowing each site to act as a backup to the other. The eventual aim was to replace the batch file transfer with an on-line, real-time update of both databases, using the 8 Mbits circuit as the interconnecting medium. The restriction on the half duplex mode of operation of local channels was bypassed by allocating two channels on each processor, one inbound, one outbound, which in conjunction with the 8 Mbits circuit provided a full duplex link.

6 Packet Switching and the Digital Connection

GENERAL BACKGROUND TO PACKET SWITCHING

Although packet switching is by no means confined to using digital technology, the combination of packet switching and high-speed digital circuits does provide the network designer with another potential solution to the design of the corporate data network. The design of a corporate *packet switching data network* (PSDN) may be based entirely around private corporate facilities, using proprietary PSDN equipment and leased point-to-point circuits. The latter are generally digital in nature to provide the speed and throughput normally required. The advantage of the private PSDN approach is that it can be incorporated with other corporate communications, such as a private voice network, enabling digital circuits to be justified on grounds both of technical capabilities and cost effectiveness.

As an alternative to a completely private PSDN, an organisation may choose to use the facilities provided by a public PSDN service provider such as a PTT, or the value added service companies that offer managed data networks. All provide high-speed digital links into their PSDNs if required. The advantage here is third party network management and maintenance, which reduces the customer's overheads for network personnel and associated equipment.

A third option is to construct a network with a combination of public and private facilities, allowing an organisation to gain the benefits offered by both types of network.

The various combinations of private or public networks, digital or analogue circuits, and single or multi-vendor supplied equipment make the potential design variations of a PSDN almost infinite. As a consequence, they allow a greater degree of flexibility to be introduced into an overall corporate PSDN design, a factor which is of benefit to many organisations.

TECHNICAL CONCEPTS BEHIND PACKET SWITCHING

The detailed concepts behind packet switching and the related CCITT recommendations are covered in a number of NCC publications, so this chapter will only touch on these areas briefly. In the main it will discuss the ideas relating to digital communications within a PSDN environment.

The Basic Principles

The basic principles behind packet switching are very similar to those behind statistical time division multiplexing (STDM). As with statistical multiplexing techniques, packet switching involves data being transmitted in the form of individually addressed blocks or data messages known as *packets*. Each packet generally contains data from a single end-user device. The amount of data is usually variable within certain sizing constraints, but the packet always contains end-user device addressing and other control information, making each packet a self-contained unit in its own right.

This approach to packet design is necessary because although each packet being transmitted across the network generally follows the same route as previous packets of that particular transmission, the PSDN could in theory route packets via different paths to their final destination point. This means that nodes within the PSDN must have access to the addressing information within the packet in order to route it correctly onwards. In all cases, the data within the packets will be delivered to the end user in the correct order that it entered the PSDN, no matter what routes individual packets took through the network.

A logical channel established between two Data Terminating Equipment (DTE) devices across the PSDN is generally known as a *virtual circuit* or more accurately, a *switched virtual circuit* (SVC). The actual act of establishing or terminating the virtual circuit is generally termed a *virtual call*. The making and breaking of virtual calls can occur with every message, transaction or session between two DTE devices, depending on the user's requirements. Another type of connection over a PSDN is a *permanent virtual circuit* (PVC). While the SVC is established

PACKET SWITCHING AND THE DIGITAL CONNECTION

only for the duration of a transaction or a session before being terminated or *cleared*, the PVC allows continuous data communication between two DTE devices without making or breaking calls between each message, transaction or session. It is in effect a permanent logical channel across the PSDN, similar to a conventional leased circuit.

The packeting/de-packeting of data and the insertion/removal of the control information in packets is generally performed by a device known as a *Packet Assembler/Disassembler* (PAD). The universal packet design standard used by PADs is the CCITT X.25 recommendation, which in its matured versions is based on the lower three layers of the OSI reference model for open systems connection. The PAD provides non-X.25 supporting DTE devices with access to a PSDN via standard CCITT interfaces. As an alternative, however, a range of DTE devices generally known as *packet terminals* are available, and these can be directly connected to a PSDN. Packet terminals were initially mainframe or minicomputers which supported X.25 emulation software and had X.25 interfaces into the PSDN. Personal computers and intelligent workstations which had options for supporting X.25 were then introduced (see Figure 6.1). Because of the power and data processing

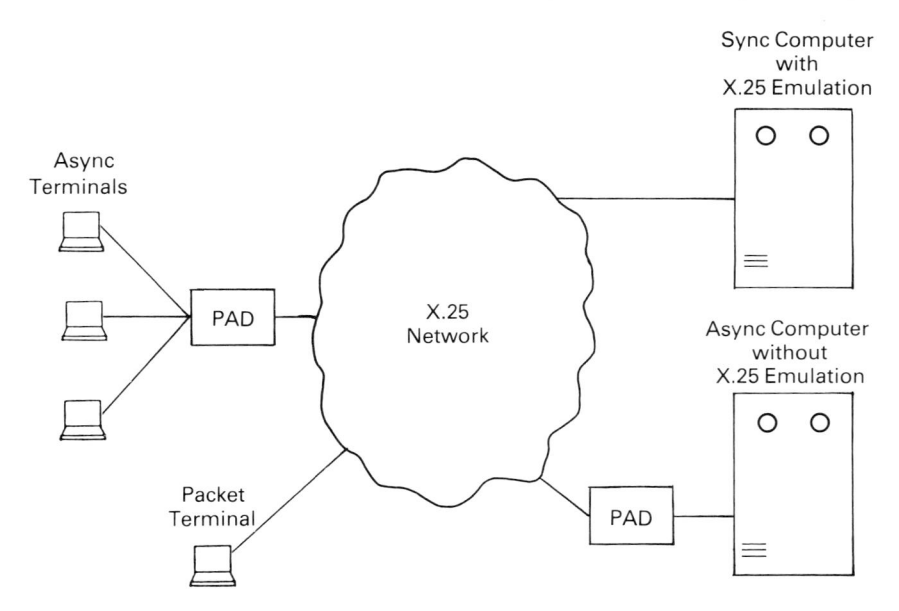

Figure 6.1 Example of a Basic X.25 Network

capability of such devices, they often require high-speed communications to enable information to be transferred quickly to and from the device, hence the use of digital circuits for such purposes.

The X.25 Recommendation

X.25 is a network access protocol, implemented between a PAD or packet-mode DTE device and the X.25 access node on a PSDN. There are subtle but significant differences between such a protocol and the more prevalent 'end-to-end' protocols, where it is the two end-user DTE devices that intercommunicate. X.25 makes no assumptions about the way in which the network functions, other than that the packets involved in an interaction between two DTE devices are delivered in the same order that they entered the network. How and where packets are routed across the network is totally transparent to the DTE devices.

X.25 basically comprises three logical layers, which are functional components taken from layers 1 to 3 of the seven-layer OSI reference model. The three X.25 layers are:

— the *physical layer*, which provides the electrical interface to the physical circuit;

— the *link control* or *link access protocol*, for error correction purposes;

— the *Packet Layer Protocol*, for addressing purposes.

The layer 1 physical interface utilises the CCITT X.21 standard for connection to digital circuits, and X.21 bis for analogue circuit connection. It is an encapsulating environment whose most important characteristics in regard to digital circuits are that it provides a bit-serial, synchronous, full duplex, point-to-point circuit interface.

The layer 2 link control is designed to provide an error-free transport mechanism for transmitting packets from the DTE to the X.25 PSDN access node. It uses the frame structure and procedures of the *High-level Data Link Control* (HDLC) protocol defined by OSI to achieve the error control and flow control required.

The layer 3 packet layer protocol defines the packet types and transitions which occur when various types of packet are transmitted or received. It does not, however, completely specify how all the control information carried by packet should be interpreted; variants are developed for particular purposes. This level also performs a multiplexing function by converting the single channel provided by layer 2 into a number of logical channels for onward connection to attached DTE devices (see Figure 6.2).

The PAD

By definition X.25 must have intelligence at both the PSDN access node and the end-user DTE device in order to support the protocol operation and handle the synchronous channel interface. Because of this it was recognised at an early stage that a large group of unintelligent asynchronous devices such as hard copy teletypes and low-cost terminals would be excluded from using the network unless provision for such devices was made. The result was the development of the *Packet Assembler/Disassembler*, more commonly known as the PAD. It is normally available in the form of a stand-alone device situated between an asynchronous device or group of devices and the physical circuit link to the PSDN. It generally provides a number of standard CCITT user interfaces, including V.24/V.28, X.21 bis and V.35. The physical circuit to the PSDN can be digital, enabling transmission speeds of up to 64 Kbits, although proprietary solutions for private corporate networks can now allow transmission speeds in excess of 100 Kbits to be obtained over digital megabit circuits.

The control and operation of the PAD is defined by three main CCITT recommendations, often referred to as the 'Triple X' protocol:

— X.3 defines a set of parameters which can be set to enable the PAD to provide various services to the attached asynchronous DTE devices.

— X.28 defines the protocol to be used between the asynchronous DTE device and the PAD. This includes call establishment, setting PAD parameters and data exchange procedures.

— X.29 defines the format of the 'supervisory' packets by which the remote end of the transmission path can signal the PAD, or vice versa. The remote end could be either another PAD or a packet terminal.

In actual operation, the PAD accepts incoming asynchronous data streams from the attached DTE devices. It then removes predefined 'packet delimiter' control characters and assembles the remaining data

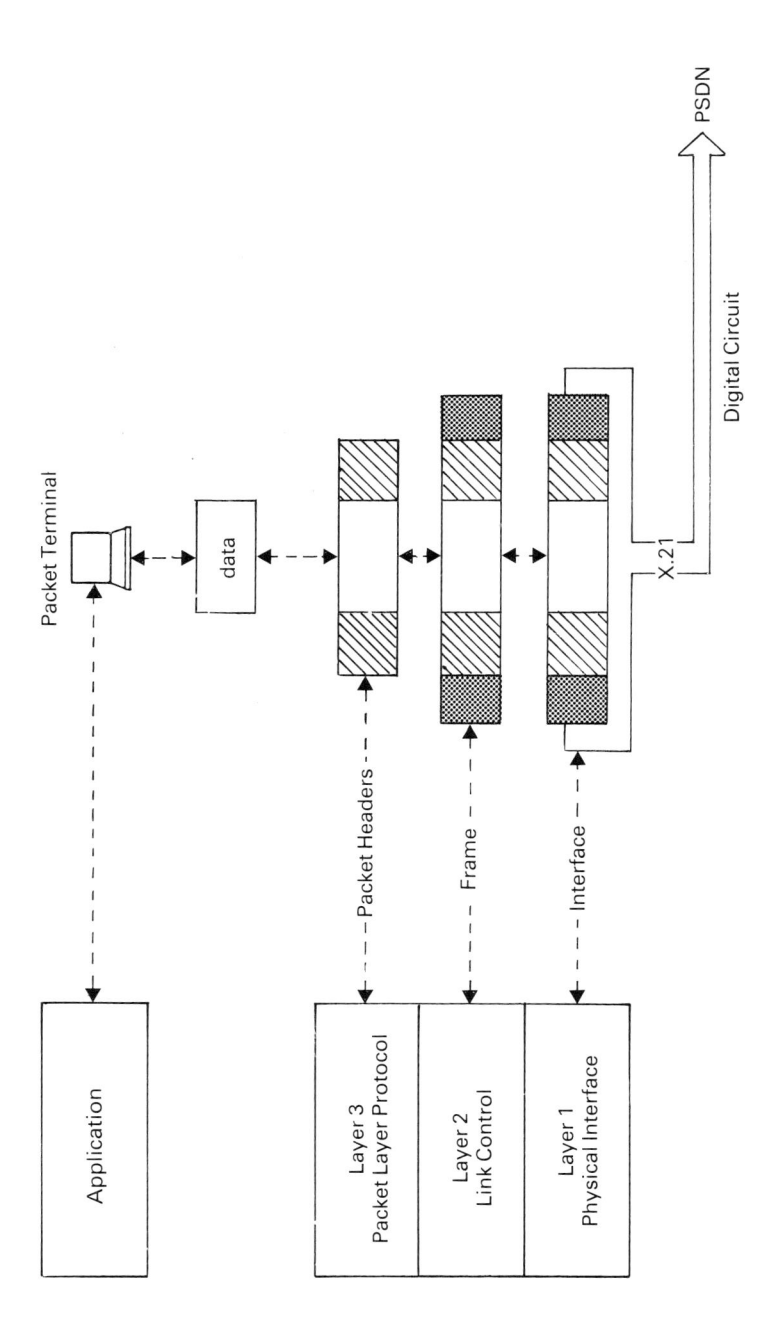

Figure 6.2 Construction of an X.25 Format Data Packet

PACKET SWITCHING AND THE DIGITAL CONNECTION 89

into packets, the packets then being dispatched across the network in accordance with the X.25 protocol. Similarly, incoming packets which are destined for an attached device are converted into character strings by the PAD before being delivered to their destination.

DTE access to the PAD can be via a number of methods. These include direct local connection by cable or LAN; remote connection via dial-up circuits over the PSTN or ISDN; or leased circuits from remote locations, these circuits being analogue or digital in nature. PADs are also available to enable support of protocols such as SNA, viewdata and telex over an X.25 network.

The PSE

At the heart of a private or public PSDN is a packet routeing device generally known as the *Packet Switching Exchange* (PSE). In its basic form it can be described as an intelligent switch to which a number of PADs, packet terminals or other PSEs could be connected. The PSE can range in physical size from a desktop device to a fully customised minicomputer. Most PSEs employ a multi-processor architecture to enable high packet switching rates to be achieved. To create larger networks, the PSE can interconnect with other PSEs, each effectively becoming a packet switching node in the PSDN. The interlinking connections between remotely sited PSEs are usually high-speed digital circuits so as to ensure good packet throughput performance (see Figure 6.3).

In corporate networks where PSEs are sited together locally, one proprietary solution is to install a Cambridge ring type LAN between the devices and use it as a high-speed 10 Mbits data bus (see Figure 6.4). In effect, this creates a large distributed PSE capable of supporting a large number of end users. Concepts applicable to data switches in general, such as automatic alternative routeing and intermediate nodes between two network access points, can all be applied to a PSE. In addition, the PSE can usually provide gateway facilities to enable its PSDN to connect to other public and private PSDNs.

It is common to find the functions of the PAD and PSE being combined into one physical device, generally known as a *switching PAD*. This provides a very versatile device, supplying facilities for asynchronous terminal concentration, host computer connection and PSDN switching, as well as supplying direct connection to high-speed digital circuits.

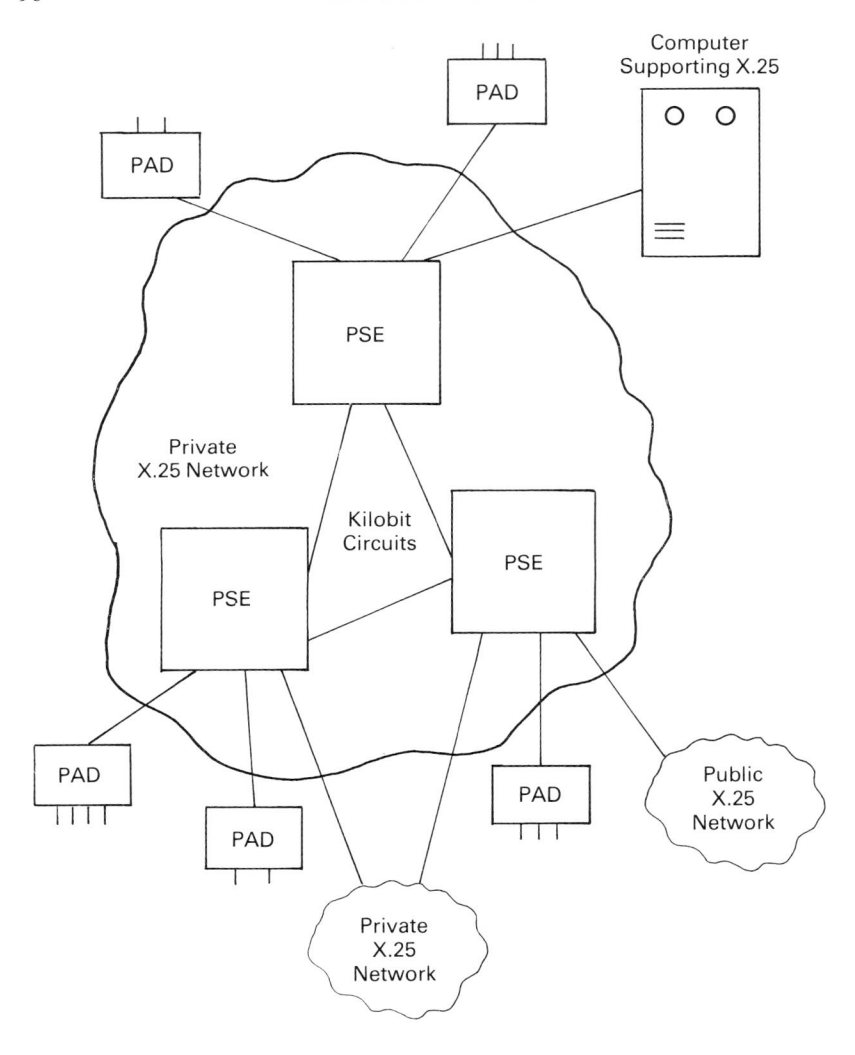

Figure 6.3 Construction of a Private X.25 Network

Packet Switching Gateways

A gateway provides a PSDN network with the ability to gain access to other forms of networks such as Local Area Networks (LANs). The gateway itself may be a stand-alone device or its functions may be incorporated into a PAD or PSE. Its purpose is to provide an interface between two networks which have incompatible higher-level protocols. In view of the large number of different forms of networks, a general purpose gateway would have to be an enormously flexible, powerful and costly device, so most gateways link to just one specific type of network: an X.25 PSDN, for instance, links to an Ethernet LAN.

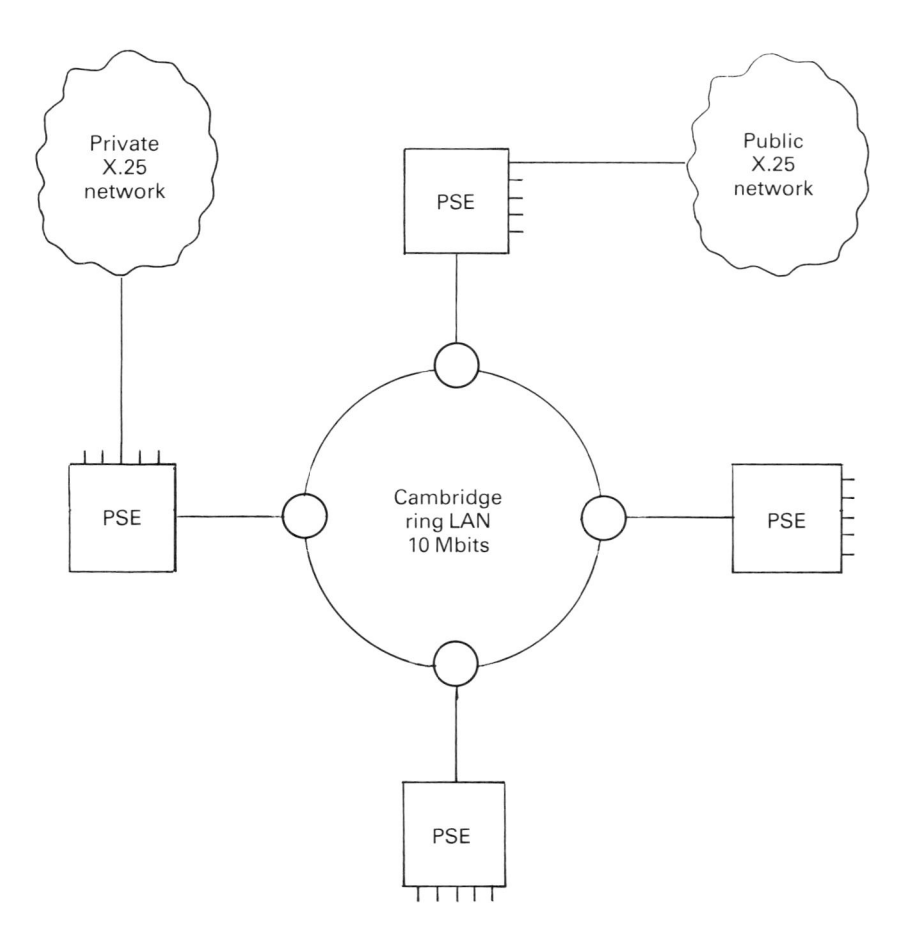

Figure 6.4 Locally Distributed PSE Utilising LAN Technology

USER ISSUES REGARDING PACKET SWITCHING

The range of PADs, PSEs and gateways available from many communications equipment manufacturers means that virtually any kind of asynchronous and synchronous DTE device can be connected to a private or public PSDN, given that the restrictions on proprietary PSDN equipment interworking have been met and that network service provider approvals have been received. The following section overviews the public and private PSDN services available in the UK today. It also covers just a few of the dozens of applications to which packet switching and digital communications could be put, showing where such a combination of these technologies could best be applied.

Public Packet Switching Data Networks

The oldest and most established public PSDN service in the UK is British Telecom's *Packet SwitchStream* (PSS). It provides a number of subscriber options including digital leased line connections between a user's premises and the network at speeds of up to 48 Kbits. The PSS provides a number of services, including a closed user group facility which basically creates a virtual private PSDN within the public network infrastructure. The 'Multiline' facility enables multiple circuits from the PSS to be terminated at one location, the circuits being delivered either by the same physical route or via different routes from the local BT access node. The facility is normally used by packet terminals such as computers to enable multiple connections to be made by remote users. It also provides resilience in case of individual circuit failure. International connections can be made from PSS to other international public PSDNs by using the *International PSS* (IPSS) service, whilst connection to other British Telecom networks such as telex, teletext and ISDN are available, as well as interworking with other UK public PSDNs such as Mercury's packet switching service.

As an alternative to a public PSDN from the PTTs, a growing number of organisations are opening up their private PSDNs for general use, enabling other organisations to use the provider's private network resources in much the same way as they would use any other public PSDN. Organisations providing this service normally run it as a sideline to their normal corporate business, but it provides a valuable source of income and allows corporate resources to be used to the full. It also enables the corporate network to be developed at a much faster rate than could have been justified if the network had remained private. Nearly all these corporate public PSDNs are digitally based, using leased kilobit and megabit digital circuits as the backbone of the service. Like the PTTs, they provide extensive network control and monitoring services, as well as 'Help Desks' where users in need of assistance can ring for help and advice.

Value Added Services

Most public PSDNs provide a number of 'value added' services, such as support of Viewdata or Videotex-type protocols. Using viewdata as an example, a subscribing organisation can obtain a comparatively cheap and effective method of supporting a number of widely dispersed, simple, dial-up terminal devices. End users simply dial into the local node of the PSDN, which then provides a fully digital route to a remote host computer supporting the viewdata database. The benefits of cheap viewdata access combined with the benefits of digital communication over long distances give a cost effective solution to the subscribing organisation, which no longer needs to set up an extensive digital network itself.

Recent surveys have shown that a number of organisations have abandoned their established corporate data networks in favour of these public PSDN offerings from other organisations and the PTTs. Many of these subscribers did so because of the cost savings involved: this type of service is charged on data volumes transmitted rather than on tariffs determined by distance. It is ideal for limited volumes of data transfer of infrequent or short duration, or where user locations are spread over a wide geographical area. A secondary saving is that the network is managed by a third party, thus saving the subscribers the overheads of investing in personnel and equipment.

There are drawbacks to these public PSDN services. A view often expressed by senior management is that direct management control is lost when a company's communications facilities are handed over to a third party service provider. Some service provider networks are viewed as being under-resourced in certain areas. Examples quoted are subscriber dial-up access nodes being saturated by calls during peak times; trunk circuits being either under-specified or over-loaded, both of which result in slow throughput; and not enough resilience being built in against network component failure. On top of this, smaller subscribers sometimes feel that preference is given to the larger users when problems or failures occur. Obviously, such potential problems have to be weighed against the potential benefits. Research and advice are essential elements in drawing up a network strategy based on third-party PSDN service providers.

Private Packet Switching Data Networks

If the volumes of data are significant, the economic alternative to a public PSDN is the installation of a private packet switching data network. Depending on the requirements, it can provide the equivalent service to its public counterpart, especially when high-speed digital circuits are incorporated into the design. Access is still available to public PSDNs if required, so the end user could have the option of accessing corporate or public services. The interconnection of the private to the public network also enables alternative routeing capability to be built into the private network infrastructure, thereby supplying a high-speed alternative path in the event of a private network failure.

One of the main reasons why a private PSDN is chosen in preference to a public equivalent is to ensure close corporate control. Control relates to the management of the network: the organisation is able to control day-to-day operation, and so in theory it can be more flexible in response to corporate requirements without having to rely on third parties. Control also relates to security. Although public PSDNs are secure, many organisations prefer the perceived additional security aspects of a private network, with its ability to further limit outside access by unauthorised users. The third aspect of control relates to cost. Since the costs of leasing circuits from the PTTs are known, fixed factors, the budget calculations can be done accordingly. No additional charges will be made on increased data volumes, as would occur with public PSDNs. Another point in favour of private PSDNs is the capability they offer for integrating corporate communications. Using high-speed digital circuits, it should be possible at the very least to combine data and voice traffic over the physical circuits between various locations. The potential for cost savings in this case is very substantial if calculated over a number of years.

There are negative aspects to private PSDNs. Due to the large number of X.25 variants, incompatibilities can sometimes occur between the various vendors' products on a private PSDN. This incompatibility problem has occurred for two main reasons. The first is that during the

1980s the CCITT produced significant revisions to the X.25 standards over and above the original 1976 recommendations. This was done so as to allow additional facilities to be introduced and to provide compatibility with the OSI reference model. The second reason for incompatibility is the subsequent interpretations made by network providers and communications equipment manufacturers of the recommendations. Some of the resulting variants have become 'de facto' standards in certain areas. This proprietary approach to X.25 was caused by two major factors: the design of the application in which X.25 was to be used; and the environment conditions such as hardware and software in which X.25 was to operate.

The network designer should be aware of these differences when designing a private X.25 network, deciding at an early stage whether to go for a proprietary solution with certain useful facilities or to use a basic version which is compatible in a multi-vendor environment.

Local Packet Switching Data Networks

When discussing local area networks (LANs), many people automatically assume the use of LAN technologies such as Ethernet or Token Ring, with their megabit operation speeds and multiple features, as the only options in local data communications. An alternative is the use of X.25 technology as a simple and cost-effective means of providing a LAN, with the additional benefit of providing direct high-speed digital communications to the outside world where required.

A 'star type' topology can be established with a central PSE connected to a number of local PADs and packet terminals, including micros and mainframes. The connections within the network could operate at up to 64 Kbits, using existing cabling where available. The network may also interface into local or remote PABXs to provide even greater flexibility. As the size of the PSDN-based LAN increases, further PSEs can be added to provide additional switching and resilience within the network.

As a footnote to using packet switching in a LAN environment, the reverse is also possible: a LAN can be connected to a wide area X.25 network. In this case proprietary PAD gateway products are available which will interlink LAN technologies such as Cambridge Ring and Ethernet to either a public or private PSDN, providing the LAN with a digital route to the outside world. In such a case the PSDN could be used as a wide area data transmission medium between two or more remotely located LANs, the medium being digitally based if required (see Figure 6.5).

The Switching PAD

There is a school of thought among some network designers which argues that the switching PAD device should provide the main access point to a

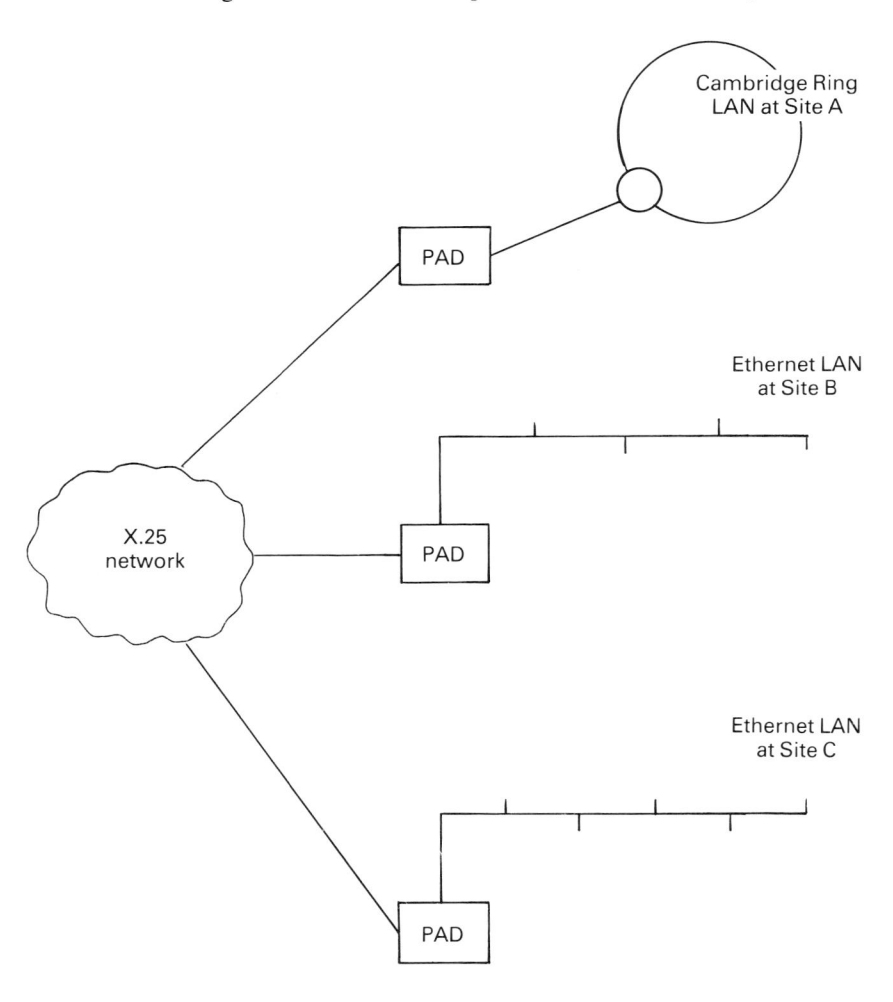

Figure 6.5   PSDN with Gateways to LANS

PACKET SWITCHING AND THE DIGITAL CONNECTION 97

PSDN, even for computers capable of supporting X.25 emulation. This type of approach to PSDN access gives a number of potential benefits:

— firstly, it may bypass any approvals problem associated with linking an X.25 emulation computer to a public PSDN;
— it eliminates the need to supply computers with emulation software or hardware, with resulting cost savings;
— it reduces the number of circuits required between DTE devices and the PSDN by enabling a large number of devices to share access to the PSDN through the switching PAD; this also means

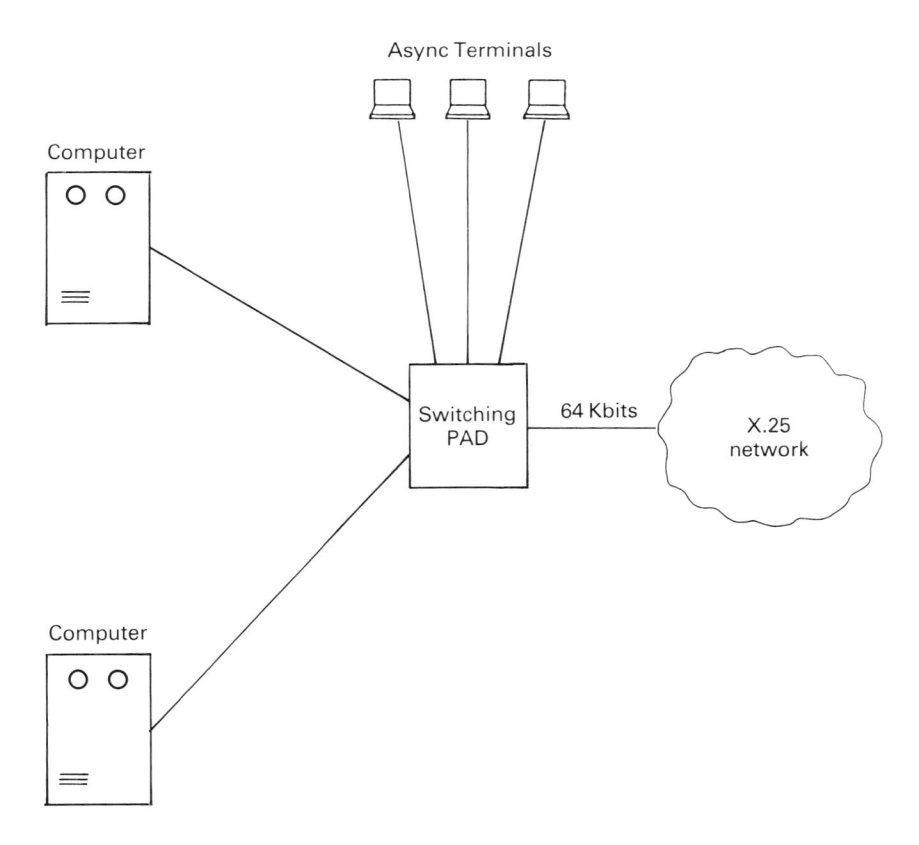

Figure 6.6 Switching PAD Providing Local and Remote Links

that it becomes more cost justifiable to use a high-speed digital circuit for the link between the PAD and PSDN;

— it provides a local switching facility for all locally attached DTE devices; this means, for example, that local terminals can connect to a local computer, or that local computers can intercommunicate, all without ever accessing the PSDN (see Figure 6.6).

In effect, the switching PAD provides a local switching solution but can also utilise the high-speed digital circuits for wide area communications.

7 LANs in a Digital Environment

GENERAL BACKGROUND TO LANs

Local Area Networks (LANs) are generally thought of as data communications systems which can be installed over limited geographical areas, and which will enable various DTE devices such as terminals, word processors and computers to intercommunicate. The two elements that make up the hardware side of LANs are the *network interface card* (NIC) and the LAN cabling.

The DTE devices are physically connected to the LAN in a fairly simple manner. Personal computers, for example, either have inbuilt NICs or expansion buses into which a NIC can be inserted. The NIC is then attached to the LAN cabling. Likewise mainframes, minis, workstations and file servers are physically connected to the cabling. The central hardware component around which the LAN is built is therefore likely to be the cabling itself, and since the cable and cable installation costs are typically up to 50 per cent of the cost of an entire LAN installation, careful planning and hardware selection is well worth the effort. Cabling types can be divided into two groups, those for *baseband* signalling systems and those for *broadband* signalling systems. Both these systems are discussed in more detail shortly, but of the two, baseband signalling is the most relevant to this publication because it uses digital signalling techniques. It is also a comparatively cheap solution to local networking needs.

On the software side of LANs the most common interface protocols are based on the standards originally defined by the Institute of Electrical and Electronic Engineers (IEEE) 802 Committee. The most prevalent today are 802.3 'Ethernet', a contention access method; 802.4 'Token Bus', a token passing protocol based on a distributed bus topology; and 802.5

'Token Ring', a token passing protocol based on a ring topology. The common element of all three protocols is that they make use of the baseband signalling system, allowing attached DTE devices to communicate across a digital pathway.

A viewpoint from which LAN technology itself can be studied is to consider the LAN as a form of digital switch. Although at first this may sound a novel concept, it is basically what most LANs do; they perform a digital bandwidth switching operation between attached DTE devices, the network itself being a form of distributed switch. My original definition of LANs stated that they are confined to limited geographical areas, but this too requires re-evaluation in certain respects. The trend today is to interlink LANs not only on a local basis, but also over wide distances. As LANs operate at megabit transmission speeds and with very low error rates, the interlinking medium must be able to provide similar speed and performance. The obvious solution is to interlink remotely located LANs by high-speed digital circuits. Product development in LAN technology has therefore produced bridges and gateways enabling LANs to interface to wide area networks. This type of integration of corporate networks is becoming increasingly important as the requirement becomes that the widest possible number of services should be available across a variety of interconnected media.

TECHNICAL CONCEPTS BEHIND LANs

Why, when and how to use LANs is a subject which has been well documented by many people. An attempt to discuss LAN technology in any great depth would require explanations in considerable detail, so instead we will discuss one small part of the subject, namely that which relates to digital technology.

LAN Signalling Techniques

Transmitting data over a physical medium such as coaxial cable requires the information to be coded in a manner suitable for that medium, as well as for the information type involved. For local area networks, two classes of signalling technique are in general use: baseband and broadband.

Baseband

Because baseband-type LANs use a digital signalling method, it is relatively simple to interface them to DTE devices which operate in a

LANs IN A DIGITAL ENVIRONMENT

similar manner, for example computers and workstations. Conversely, it has proved difficult and expensive to connect non-digital devices such as telephony equipment. Baseband signalling involves no modulation in the normally accepted sense of the word. Instead it employs discrete changes in the signal level to represent the binary information content of the transmitted data.

One of the most widely accepted ways of encoding the signals onto the medium is 'Manchester encoding'. Manchester encoding is one of the simplest methods to implement and has a built-in clocking scheme which enables every system on the network to remain in synchronisation. To explain the Manchester encoding technique, a sample data stream using Manchester encoding is shown in Figure 7.1. Each time interval is divided up into equal cells, each of which is used to represent a single bit. Each cell is then in turn divided into two, with the signal level in the first half representing the complement of the bit value being sent in the cell. In the second half of the cell, the uncomplement value is sent. In this way there is always a signal change within a cell at the halfway point, which ensures that devices can be kept in synchronisation without the necessity for separate synchronisation signals.

Broadband

Unlike the digital signalling method used in baseband, broadband LANs are purely analogue in nature. To draw the distinction between baseband and broadband, the latter is briefly outlined below.

The broadband technique has its origins in the cable television

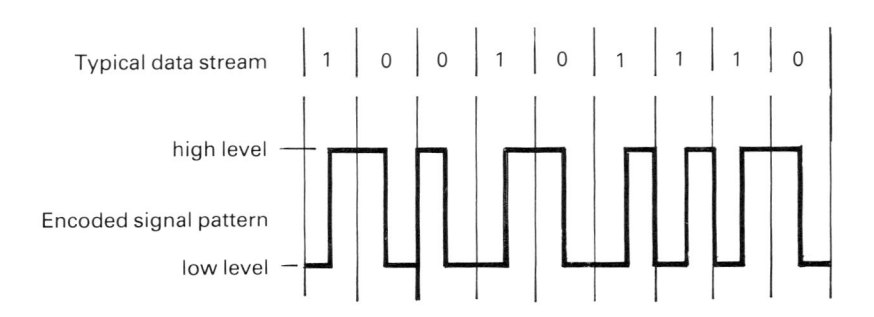

Figure 7.1   Manchester Encoding

industry, its most significant user base being in the USA. Essentially it is a method of 'frequency multiplexing' many user channels onto a single physical cable. Typically a broadband system has a bandwidth of 300 MHz (300 million cycles per second), so assuming that a speech range of 300 Hz to 3400 Hz is rationalised to a bandwidth of 4000 Hz, then a 300 MHz system could in theory handle 75,000 voice channels (ie 300 MHz ÷ 4000 Hz = 75,000 channels). In actual usage, the system could handle a combination of voice, data, text and image transmissions in any combination of numbers or bandwidths. Broadband systems use a one or two cable implementation to link all the devices to the network. The actual interface between the device and the cable is performed by a radio frequency transmitter/receiver unit termed an 'RF modem'.

To summarise the two signalling techniques, baseband uses a digital signalling method, is simple to install, virtually maintenance free, and can utilise several cable types, including twisted pair and coaxial. Broadband systems use an analogue signalling method, are usually more costly to install than baseband and require regular maintenance. Their main advantage is the exceptionally wide bandwidth, which can be split into multiple channels supporting various communication techniques.

USER ISSUES REGARDING LANs

The rapid developments in LAN technology are opening many new application areas. Many of these developments have involved digital technology, a few of the more important advancements being outlined below.

LAN to WAN

Recent innovations in LAN application technology have been to interlink LANs into larger networks, the individual LANs being located either locally to each other, or spread over wide geographical areas. In this latter case, the creation of a *Wide Area Network* (WAN) has usually involved some form of high-speed link to interconnect the LANs, one of the more obvious choices for this link being the leased digital circuit from the PTTs.

Megabit Connection

One of the first products in the area of high-speed LAN to WAN interconnection was developed and marketed by the Network Systems

LANs IN A DIGITAL ENVIRONMENT

Corporation of the USA. The local area network part of the system is a product called Hyperbus, which uses a baseband bit-serial bus to transport data over a coaxial cable at up to 10 Mbits per second. Other LAN cabling systems could be used if required. Personal computers, workstations and other similar DTE devices can be attached to the LAN by a microprocessor-based NIC known as a *Bus Interface Unit* (BIU). These intelligent transceivers take data from the attached device and translate it into a speed and message structure suitable for the receiving device. These messages are then sent either to other devices on the LAN or out onto the high-speed digital circuit using the wide area part of the system, called 'Hyperchannel'. The interface between LAN and WAN is provided by a gateway BIU.

Hyperchannel was originally developed to interlink dissimilar computer processors and peripherals over a coaxial cable at speeds of up to 50 Mbits per second. Its evolution enabled it to utilise the then emerging high-speed digital circuits of the PTTs. The first long-distance UK installations of the system used the 2 Mbits circuit options although increasing use is being made of 8 Mbits circuits. The combination of high-speed LAN and WAN provides an organisation with the option of interlinking many types of processor and other DTE devices at high transmission speeds both locally and over digital circuits. The disadvantage may be the capital outlay for such a system, making it a worthwhile proposition only for larger organisations.

Kilobit Connection

Several proprietary LANs support the ability to provide access to one or more high-speed digital circuits operating at speeds of up to 64 Kbits. This enables either high-speed access for remotely sited DTE devices or provides a link between two or more remotely located LANs. Access to the LAN is generally provided by a proprietary gateway device developed specifically as a WAN gateway. It links to the LAN and provides a standard CCITT V.24/V.28, V.35 or similar interface to the end user. This is generally directly connected to the NTU/CTU of a digital circuit. Some proprietary LANs allow the clocking signal from within the PTT digital network to be used as the primary clock source for the LAN, bypassing the LAN's own internal clocking source. This ability obviously provides better communications synchronisation between the two networks, or between the PTT network and several remotely connected LANs.

PSDN Connection

Another LAN to WAN interconnection method was discussed in Chapter 6, namely LAN connection to a wide area packet switching data network (PSDN). Proprietary PAD gateway devices can interlink LAN technologies such as Cambridge Ring and Ethernet to either a public or private PSDN, the link operating at up to 64 Kbits over digital circuits. The ability to connect LANs to PSDNs means that remotely sited LANs can intercommunicate over long distances, the interconnecting route being fully digital if so required. This method provides a cost-effective way of establishing LAN communications at a fairly high data transfer rate, and is especially useful for organisations without extensive wide-area corporate networks that could link the gateway devices to a public PSDN.

Digitised Voice Over LANs

Until recently, the integration of voice communications onto a LAN has been one of the main stumbling blocks to an integrated digital local network. Compared with data or text, a voice service is relatively insensitive to errors, but it is particularly sensitive to transmission delay. This fact can easily be shown in any telephone conversation. Although the person listening can usually understand the meaning of a message when the words are distorted or even lost, the listener might well find the conversation difficult to follow if it was broken down into small sections, with significant gaps in between. Even worse would be the case where individual words were punctuated with gaps. Even so, the potential advantages of integrating voice onto a LAN has resulted in a lot of time and money being invested in 'voice over LAN' research by the communications manufacturers, particularly if it could be offered as an additional option to existing baseband LAN systems already available.

Converting voice signals to digital form has usually involved using PCM techniques to produce a 64 Kbits quality voice channel. To reduce this bandwidth overhead on the overall LAN throughput ability, alternative voice coding and sampling systems have been tested. These tests have shown that it is possible to reduce the bandwidth down to 16 Kbits and still retain adequate voice quality. Of all the currently available LAN techniques, a packet mode LAN is seen as one of the best options for carrying the digitised voice signal. With a packet mode LAN, the voice signal (like the normal data signal) is broken down into discrete messages or packets before being placed on the LAN and transmitted. In order to reduce packet transmission overheads to a minimum, a silence suppression technique can be introduced at the transmitting point: this eliminates the 'empty' packets which occur when a person is silent.

Using the variety of techniques available, the 'voice over LAN' results look encouraging. The main problems experienced are transmisson delays caused by the voice sampling techniques and packet queueing. In these cases, the round trip delay can be well above the maximum 2 milliseconds delay currently specified for digital PABXs. In the short term, the most useful application for the technique in the context of LANs might be to interface the 'voice over LAN' technique with a centralised voice messaging system. In this manner a fragmented conversation would not matter so much, as the voice messaging system would automatically concatenate the message. The system could also interface to the wide-area corporate digital network.

The drawbacks to 'voice over LAN' communication will eventually be resolved, especially with the introduction of broadband LANs operating at speeds of up to 100 Mbits or above. This type of LAN will be able to support large numbers of both voice and data channels, and will become, in effect, a digital telephone system in its own right. An example of high-speed LAN technology can be found in the ANSI Fibre Distributed Data Interface (FDDI) standard, which specifies a 100 Mbits token passing physical ring using fibre optic cable. FDDI-II is a subsequent companion standard of FDDI, and supports both packet switching and circuit switching modes of operation. FDDI can provide high-speed local connection of computer processors and peripherals, together with 'backbone' support for multiple medium peformance LANs. FDDI-II can in addition provide support for voice, and therefore together, FDDI technology can provide the basis for an Integrated Services Local Network (ISLN), a local and complementary version of the ISDN.

8 Digital Network Control and Management

GENERAL BACKGROUND TO THE SOFTWARE SOLUTION

The emergence of digital technology within the communications world has coincided with a move by the communications equipment manufacturers away from purely hardware-based solutions to a combination of both hardware and software-based devices. The result has been a series of new developments within two fundamentally important areas of communications technology.

The first has been the redesign of the physical communications node, with the development of a new generation of hybrid devices which are basically hardware shells with plug-in intelligent software modules. The move towards the hybrid communications device, with its combination of hardware and software elements, is important to the manufacturer because it provides greater flexibility in the design and development of new products, software being the key to future trends. These products tend by their very nature to be more 'future-proof', being easily field upgradable with the simple addition of new software modules. This gives the hardware a greater lifespan and the capacity to be adapted to a number of uses. The result is a device which tends to 'lock in' existing customers to the vendor's product range. This is not necessarily bad news for the customer: the initial investment in such devices is fairly secure given that the equipment has an evolutionary path rather than a 'cul-de-sac' approach to its design. This ability to evolve existing installed equipment is particularly relevant to the comparatively new area of digital communications, which is still undergoing change that will affect the final structure of any corporate digital network.

The second key area of software development has been in software-based network management and control systems. These employ the latest technologies in such areas as integrated databases and colour graphic displays to provide a set of tools that can effectively manage and control today's increasingly complex corporate networks. These tools are applicable to both local area and wide area networking, particularly the latter where the intelligence may be of a distributed nature, spread over a wide geographical area. To coordinate such distributed intelligence, some form of centralised control is mandatory in order to allow network operators to interact with the software and provide overall guidance to the whole network as well as individual network components.

Although most communications manufacturers tend to provide network control and management software either as a single or an integrated package of systems, it is useful to discuss these two functions as individual activities. The network control function will comprise the activities relating to the actual day-to-day operation of the network, such as real-time monitoring, alarm and fault detection and actioning of network operator commands. The network management function will cover the longer-term strategic issues such as network performance, capacity and planning.

Another view of network control and management systems is to consider these systems in a hierarchical structure, with a number of distinct levels. At the bottom level are the individual network components, such as multiplexers or data switches. Each of these is self-contained with its own configuration and control functions (now more commonly software based) which allow control to be exercised remotely if so required. Above this level there is generally the proprietary *Network Management System* (NMS), which will control and coordinate a number of similar proprietary network components. The NMS generally resides on a micro or minicomputer attached to the network. At the next level up, in addition to or instead of the proprietary NMS, there may be some form of 'wrap-around' monitoring system, again a proprietary solution to allow some level of management to take place on a multi-vendor network. A new level at the top of this structure could be a bespoke network management coordination system. This type of system has in most cases been specially developed by individual user organisations to gather and coordinate information about the network being generated by the lower level systems, as well as from other sources such as attached computers and FEPs. It also allows input from the network operators themselves.

This description of a hierarchical control and management structure is

DIGITAL NETWORK CONTROL AND MANAGEMENT

for conceptual purposes only, but it does illustrate the varying levels of complexity to which software based control and management may be taken (see Figure 8.1). However, such control and management systems are vital, especially considering the growing complexity of digital networks. They are particularly important for organisations that wish to maximise the return on their investment in digital communications technology.

At the centre of most control and management systems today is the network database. It receives, stores and allows access to information that is important both to the day-to-day control and to the longer term management activities of corporate network supervision. This generally

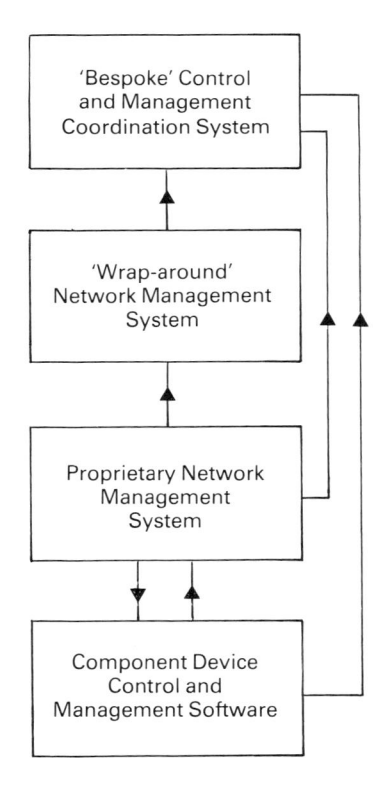

Figure 8.1 The Conceptual Hierarchical Structure of Network Control and Management Software

includes details on network configuration, faults and potential failures, performance figures and many other varied facts. The database may also allow storage of other data such as network administration details, equipment inventory or other user information. To access such a potentially vast collection of data, the control and management system must include a comprehensive set of manipulatory tools which will allow the network operator or designer to steer through the database and collect the required information.

To present the results from software based control and management systems in the most concise form possible, most communications manufacturers have opted for a colour graphics display as the output method to supervisory terminals or PCs. This was initially seen as an expensive novelty by many corporate network managers, and treated as such. Their opinion changed as it was found that colour allows fault identification, network topology diagrams and textual information to be interpreted far more easily than was possible on a purely monochrome screen. For the software designer, the potential use of 32 or more colours means that far more information can be contained in a single character position than could be achieved in a textual type of display.

Most control and management systems allow additional supervisory peripherals to be attached in addition to the primary control terminal. Enquiries and results can be undertaken by other terminals on the network, with output directed to a hard copy printer. Output results can also be directed to a database for subsequent retrieval or interrogation.

TECHNICAL CONCEPTS BEHIND CONTROL AND MANAGEMENT

Network Control Features

The primary objective of any network control software must be to maintain the maximum operating efficiency of the network, minimise network downtime, and generally ensure that an acceptable level of service is provided to the users on a day-to-day basis. To achieve this, the network control software requires a large number of control commands, comprehensive diagnostic tools and good network configuration facilities. These features can be broken down into several key groupings:

Configuration: the set up and modification of the network topology, either on an off-line or real-time basis;

DIGITAL NETWORK CONTROL AND MANAGEMENT

Command and control: interrogate the network and perform real-time changes when necessary;

Fault detection and recovery: detect faults and automatically take recovery actions if applicable;

Monitoring and journalisation: gathering and recording events on the network.

These features are now discussed in slightly more detail, together with their relevance to digital networks.

Configuration

This gives a network operator the ability to create, update and delete the software parameter settings of an individual communications component, a group of components, or all components within a corporate network. This function is usually achieved by the use of inbuilt intelligence software via some form of component control panel, or a suitably attached supervisory terminal or PC.

The intelligent software will lead the operator through a series of parameter options, generally displayed in menu form. These allow the software (and the operator) to gradually build up what is generally termed a 'configuration map' or 'table'. Some of the more advanced systems allow experienced operators to enter the configuration map language statements directly, bypassing the menu screens. Other systems allow the operator to duplicate sets of parameters several times, or to copy in portions of other maps held elsewhere. Both features speed up the time taken to build configuration maps.

Once the system has validated the configuration map as being syntactically correct, it is useful to be able to store the map for future reference. This is usually achieved either by writing it into non-volatile memory or by storing it on some form of hard disk or diskette. Either way the map can be recovered even after a power failure, an important consideration when recovery delays after network failure must be kept to a minimum. Storage also allows a number of configuration maps to be built, stored and used on different occasions when needed. When required, the configuration map can be loaded into the designated communications component. This could be achieved either by manually loading a diskette or (more usually) by using telecommunications facilities to download the map from its storage place to a local or remote target device or devices. The new map will then replace any existing configuration when the command is issued to reload or reinitialise the device.

Configuration map loading and execution procedure is normally a manual procedure carried out by the network operator with the use of the controlling software. However, most systems will have automatic reconfigure options. For example, at predefined times the system will automatically load predefined maps, changing the topology of a network from a daytime user terminal configuration to a night-time computer file transfer configuration, thereby optimising the network for a particular user application. Automatic reconfiguration could also take place on the activation of an alarm condition, so that the network could automatically reconfigure in the event of circuit or component failure. Because loading and executing a new configuration map normally involves some disruption to anyone using the network at the time, reconfiguration is not normally undertaken during peak usage hours. However, most systems have the option to alter individual parameters within the existing configuration map of a communications component without disrupting the traffic flow. This allows the network operator to 'fine tune' a device while users suffer only a slight inconvenience. In most cases this would be a temporary loss of transmission while the parameter alteration was undertaken.

Obviously, all the configuration features described here are particularly important to devices in a high-speed digital network. This is because digital networks tend to support larger numbers of users and applications than their analogue network counterparts. The configuration maps therefore tend to be far more complex and more numerous, particularly when the topology of the network may well change several times during any 24-hour working period. The need for flexibility in a digital network is important, and this is exactly what software-based configuration methods can offer.

Command and Control

This feature allows the network operator to interrogate and control the network control system, by entering predefined commands and control statements or initiating the execution of programs via the supervisory terminal interface. Once entered, the commands should be interpreted and actioned by the control software and the results displayed to the

DIGITAL NETWORK CONTROL AND MANAGEMENT

network operator. Such a feature allows 'human judgement' to be employed in the overall network control function, thus making the network as flexible as possible to meet the needs of the end users.

The command functions available to network operators depend largely on the type of proprietary system used, but as a general guide they can normally be categorised as follows:

Parameter Modification: The ability to inspect and modify software parameters within communications components on the network. This could include the addition/deletion of channels, alteration of data transmission rates or the setting up of network access passwords.

Diagnostics: The ability to initiate and perform diagnostic routines and present the network operator with the results. This includes initiating 'test and compare' routines, where test patterns are sent around a defined route in the network.

Database Access: The ability to access and manipulate information held on the network's database, and present it to the network operator in a variety of forms.

Threshold Levels: The ability to establish error threshold levels for each component in the network. This means that if the number of errors occurring on a defined parameter exceeds the preset threshold value, an alarm condition occurs and appropriate actions are initiated.

Priority Setup: As with any resource, there is a maximum level at which the resource can be utilised, so priorities can be established for the uses of such a resource. In practice this means establishing priorities for users, channels, device interfaces and other similar groups.

Remote Switching: Some devices or subscribers on the network may not have the ability to route themselves around the network. This function is therefore carried out by the network operator.

As with configuration, the command functions available to the network operator are particularly important to the effective control of a digital network. What makes these command functions so important is the potentially large number of users who could be affected by poor network performance.

Fault Detection and Recovery

Although this is technically part of the monitoring and journalisation routine taking place continuously throughout the network, fault detection involves the ability to detect exceptional conditions occurring within a particular network component, and then take appropriate action. Once a problem is detected, the system should notify the supervisory terminal of the fault, activate any alarm condition if the condition has been set, and undertake automatic correction and recovery procedures if any such procedures have been specified. This could take the form of attempting to restart a failed component, or initiating automatic rerouteing of channels to bypass a failed circuit or node. At the same time the network operator should be able to track down the fault via the supervisory terminal, undertake additional manual diagnostic routines if required, and start taking appropriate action to correct the fault.

On a digital network, a failed component or circuit could cause inconvenience to a large number of users, so automatic fault detection and recovery is vital to ensure minimum downtime. To manually reroute a large number of users on a megabit circuit would be a slow and time-consuming process, even using a supervisory terminal. With careful design and planning, the system should be able to reroute automatically in a fraction of the time it would have taken for a human to respond. In the meantime the network operator is left free to carry out the fault location and correction task. This is a classic example of how an NMS and a network operator can work in parallel on the same problem. Each tackles it from a different angle, and each is applied to the most appropriate area, thereby providing a better response to the users' needs.

Monitoring and Journalisation

This is the ability to collect and store information relating to the status of individual components within the network, as well as the overall operation of the network itself. It is useful not only for subsequent processing and analysis for network management functions, but as an important part of the day-to-day network control system. Such information is useful for immediate retrieval via the supervisory terminal in a number of forms and displays. It can provide statistics on events which have occurred over the immediate or recent past and so provide the network operator confronted by a problem with vital information. For example, the current statistical record of input/output errors on a device can be compared with those of the previous day, week or month and

DIGITAL NETWORK CONTROL AND MANAGEMENT

thereby provide a clue to a current fault on the component.

Network Management Features

Network control software provides the network operator with day-to-day operational control of the network. Network management software should provide the ability to manage and analyse the network, furnishing the network designer with the information to optimise resources and network performance, monitor current and potential trends, and generally allow for a planned and ordered evolution of the network to take place. The facilities within network management software should cover the following areas:

Analysis of statistics: the ability to produce graphs, reports or other listings to show usage, throughput, errors and other statistics on any component monitored by the software.

Performance monitoring and prediction: the ability, using such statistics and with the aid of other software tools, to produce component and network performance and prediction forecasts.

Network capacity planning: the ability to supply network designers with information which will help them to develop future network designs.

Change management: the ability to provide a managed environment for all changes taking place on the network.

A point to bear in mind about network management functions is that their existence or otherwise should not impair the day-to-day performance of the network. In practical terms this means that if the network management software failed, it would not affect the network control software or the network itself.

Analysis of Statistics

The information originally collected and used for the network control function of on-line, real-time monitoring of the network can also normally be stored and subsequently used as data to produce statistics on the functioning of the network. These could take the form of graphs, diagrams, reports or other listings such as:

Channel usage: Information such as total characters transmitted and received; percentage of data compression attained; and the rate of transmission errors. These figures can all provide the base data with which network optimisation and tuning can be undertaken.

Device statistics: As with channel statistics, details on any network component such as a multiplexer or switch can be obtained. Statistics might include peak character throughput per second; buffer usage; and hardware faults. These would allow analysis to be made of each component.

Alarm log: A chronological report of any or selective alarm conditions which have occurred over a defined time period.

It should be possible to gather the collection and journalisation of all such information either on a selective basis or in total, depending on what information is required and over what time period it is to be analysed.

The ability to be selective is important. Even on medium sized networks it is quite possible to gather large volumes of performance and operational data, relating both to the network and the users, in a short period of time. This can put strain on both storing and processing such information. When processing such data, it is vital that good software tools exist to manipulate the data into a useful format, or it remains a collection of meaningless figures. Once produced, the statistics should enable the network designer to undertake the remaining tasks outlined earlier.

Performance Monitoring and Prediction

In many cases, network components and transmission circuits deteriorate over time rather than failing instantaneously. The monitoring of the network should provide information which points to such trends. It would then be possible to undertake preventive maintenance, and so provide a better uptime percentage than would have otherwise been possible.

Network Capacity Planning

As all networks are subject to performance level requirements, and other applications are constantly being developed for use over the network, it is important that the network designer has the ability to predict the current and future capacity available to the network. The network management statistics should provide the information on which such predictions can be

DIGITAL NETWORK CONTROL AND MANAGEMENT

made, and therefore allow a smooth and cost-effective evolution of the network to take place.

Change Management

The testing and release of new network software should be a function tightly controlled by the network operator. It should be part of the network management and control system that part or the whole of the network software and/or the configuration map could be tested in the live environment during non-critical periods, before returning back to the current versions for live operations.

The Network Database

Both network control and network management tend to be based more and more on the idea that a database of network information is central to all functions. In the case of network control it provides for the immediate needs of the network operator, and for network management it provides the base information from which network designers can extract future requirements. In addition to the many uses already discussed, the database can also be used for storing and displaying other network details. These can be summarised as:

Administrative

The creation and updating of an administration database allows details such as site locations and telephone numbers, numbers and types of terminal equipment, FEP channel numbers or similar information to be stored. It can be referenced either when viewing network schematics (where such information could be overlaid on the supervisory terminal) or, when sorted, in the form of listings and reports. The database could also collect statistics relating to network usage for subsequent billing purposes.

Inventory

An inventory database is useful for organisations with large amounts of equipment which by its nature tends to be moved around the network from time to time. This type of control allows a trace to be kept and provides an accurate audit ability when required.

User Information

The user information database allows various details such as the logging of failed circuits and equipment to be undertaken. Details of the fault (eg when the fault occurred, service engineer response time and when the fault was corrected) would allow the network operator to monitor the number of faults occurring together with engineer response times to call-outs. It could also be used to predict time intervals between faults on defined equipment and therefore allow a better preventive maintenance schedule to be arranged.

USER ISSUES REGARDING NETWORK CONTROL AND MANAGEMENT

The Network Management System

As I originally stated, control and management software is usually supplied as an integrated package generically termed a *network management system* (NMS). This is generally proprietary in nature, and is usually developed by the communications equipment manufacturers. These proprietary systems tend to operate on stand-alone processors such as PCs or minicomputers, with communication links into the network and the associated proprietary network components. Alternatively, some NMSs reside on one or more of the network components themselves, exercising control of similar devices within the network. Network management systems can generally be grouped in three categories: 'proprietary equipment only' network management systems; 'wrap-around' network monitoring systems; and specific bespoke network coordination systems.

Proprietary Network Management Systems

If the corporate network hardware and software components are from one particular vendor, then in most cases that vendor will supply a network management system to control and coordinate the network. This will comprise some or all of the features discussed earlier in this chapter. In many cases, the NMS will provide the network operator with a greater range of control and management features than was available on the basic control software on individual devices. Some proprietary systems are better than others, but in most cases the choice of NMS will be governed by the existing network components. However, if the network is being built or rebuilt then both hardware and software are open to re-evaluation.

The drawback with any network management system is that is a proprietary solution only, so that networks comprising several vendors' products will prove difficult if not impossible to manage effectively. Several NMSs could be employed, but this would result in multiple supervisory terminals and possible communication problems due to the lack of interworking between proprietary systems. With the advent of Open Systems Interconnection for attached devices, the real need is for this approach to be applied by communications equipment manufacturers at the network control and management level. Unfortunately any stable international standards in this area are still some years in the future. At the moment the only clear alternative to the proprietary NMS is the 'wrap-around' NMS, which would replace or sit above one or more vendor-supplied NMSs.

'Wrap-around' Network Monitoring Systems

The 'wrap-around' network monitoring system allows the network operator to achieve a level of network monitoring and management over a multi-vendor equipped network. On a digital data network, the system works by placing a separate monitoring device at remote circuit termination points, inserting the device between the customer interface side of the NTU and the DCE side of any attached user equipment. At the local end of the digital circuit, a similar monitoring device is inserted between the NTU/user equipment interface. This monitor is then linked directly to the central NMS processor (see Figure 8.2).

Once the monitoring devices are in place at both ends of the digital circuit, they can intercommunicate using an inbuilt multiplexing feature which allows their network management information to be added to the existing data traffic. This combined data stream is then carried across the circuit and demultiplexed by the receiving monitor. In the case of the local monitor, the information is then passed to the central NMS processor for analysis. The central NMS can thus receive status information about the digital circuit from both the local and remote endpoints. It can also instruct the monitors to undertake tests, such as setting up local and remote loopbacks and other similar routines.

The advantage of the 'wrap-around' NMS is that it can be used in a

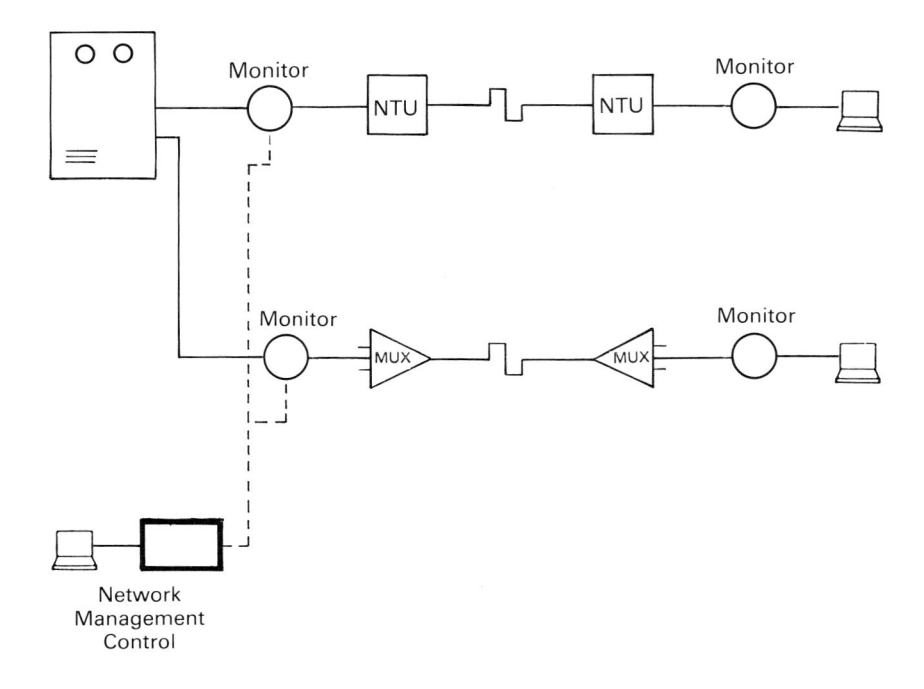

Figure 8.2 Example of a 'Wrap-around' Network Management System

multi-vendor environment. The disadvantage is that it provides monitoring functions only, although some simple test and diagnostic features are available to the network operator. Control of other devices such as multiplexers and switches is outside the capability of such a system, so other proprietary device management systems may be required. Another negative aspect is that the monitor device is yet another 'link in the chain', and therefore another potential point of failure in the network.

Bespoke Network Coordination Systems

As well as the information provided by the NMS and other network component control and management software, information about the network's status and performance may also be created by attached devices such as corporate mainframes, FEPs and the like. This could lead to an 'overabundance' of performance, event and alarm messages being generated by several systems, each with its own particular view of the network.

DIGITAL NETWORK CONTROL AND MANAGEMENT 121

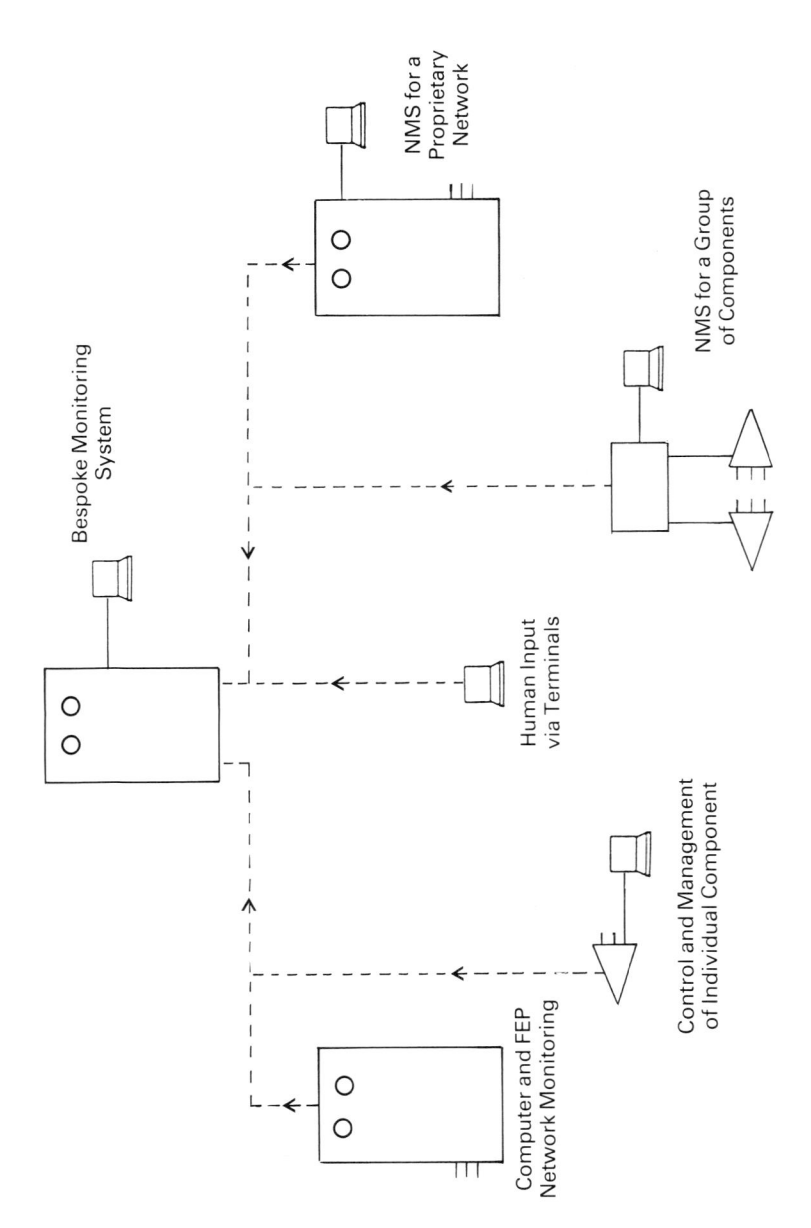

Figure 8.3 Example of a Bespoke Network Monitoring System Interfacing to Other NMS Systems

To prevent confusion and duplication of effort by different DP groups within an organisation, some form of coordination is required. In some organisations this has taken the form of developing a 'bespoke' software system running on a micro or minicomputer which has input from various components via a direct communications link, or via terminals where information is entered by computer operators, network operators and other associated groups (see Figure 8.3).

The basic idea is to cross-reference events from various input sources and provide an overall picture of the status of the network, but in a condensed form. This can then be used to pick out key facts from the mass of input data, allowing network operators to improve response times to any problems.

Other Benefits of a Network Management System

There are many advantages in the use of some form of network management system, be it a simple system or a complex one. Apart from some of the benefits already discussed, an NMS also offers:

— The ability to record and report events on an individual digital link. This gives the network operator hard copy evidence when discussing circuit problems with PTT engineers. In general, such reports are taken as fairly conclusive evidence when intermittent faults are occurring on a digital circuit, and therefore help to resolve such problems more quickly.

— NMS provides a way of concentrating staff at key locations as well as allowing less experienced staff to control the network. Skilled staff are a rare and expensive resource.

— Diagnostic facilities. With such facilities available through the NMS, the requirement for additional diagnostic and test equipment may be drastically reduced, if not eliminated, with appropriate cost savings as a result.

In a large digital network, or indeed a large analogue one, the ability to control and monitor is vital; failure could cause great disruption to a large number of users. A network management system should allow the network operator to pinpoint quickly any fault on the network. In effect, it should generally provide the network operator with the means to manage the network, rather than the network controlling the operator.

9 The Digital PABX

GENERAL BACKGROUND TO THE PABX

The evolutionary development of the *Private Automatic Branch Exchange* (PABX) has been gaining momentum over the last decade or so, and during the last few years a number of new and important innovations have been introduced. Of all these events, the migration from analogue-based to digitally based technology within the PABX must be regarded as one of the major advances in PABX design. Although the driving force behind this move was mainly due to the cost savings involved both for the manufacturer and the customer, it also enabled a number of new features and services to be introduced.

The main drawback in the early years of the digital PABX was the fact that the PTT public networks with which is was to interwork were analogue-based. For two digital PABXs communicating across the PTT analogue network, this meant an analogue/digital conversion would have to take place at least four times:

— between the analogue channel from the telephone handset to the digital PABX;

— from the digital PABX to the analogue network;

— from the analogue network to the remote digital PABX;

— from the remote digital PABX to the remote analogue handset.

This repeated conversion process has an impact on the quality of the signal, distortion being the obvious result.

The advent of digital communications supplied by the PTTs has enabled the PABX manufacturers to develop and enhance the PABX functionality so that it can make the best use of this medium. As discussed earlier (in Chapter 4, Digital Data Switching) the digital PABX is capable of providing switching and other services for data communication, but its primary application now and for the immediate future is still in handling voice transmission, with all the features and facilities expected from the modern digital telephone exchange. These include 'auto callback', the 'follow me' feature for call redirection and 'short code' dialling.

Digital PABX features and services can be extended across the private digital corporate networks, interworking with other PABXs sited either locally or remotely to provide an integrated corporate communications system. In addition, integration of private networks into public ISDNs will eventually enable organisations to utilise digital PABX services worldwide. This realisation of the advantages of digital PABXs has become more evident in many organisations over recent years. As an illustration of this, in the mid-1980s there were approximately 40,000 PABXs with 100 plus extensions in the seven major countries of western Europe. Of these, only one in seven was of digital design. Within five years, although the total number of installed exchanges has changed very little, almost every new PABX installed will be digital, changing the overall ratio of digital PABXs to four in every seven.

As I stated earlier, the immediate and short term acquisition policies of PABXs by most organisations will still be primarily dictated by voice requirements. Increasingly, however, a growing secondary requirement is becoming evident: that the device should offer integrated digital switching to support voice, data, text and image. The prime requisite is that the PABX should be flexible enough to exploit future developments in digital technology as and when they arrive.

TECHNICAL CONCEPTS BEHIND DIGITAL PABXs

For those who are not familiar with the design and use of PABXs, there are several publications available from the NCC which cover this subject in varying degrees of detail. This section will therefore cover the aspects of digital PABXs which relate to their use with digital communications, and the resulting facilities and services this provides.

Digital Signalling Systems

Two important digital signalling systems exist within the UK to enable the digital PABX to interconnect with the outside world. Both cover separate but important parts of the UK's digital network and both form the basis for the UK's Integrated Services Digital Network (ISDN).

DASS2

The *Digital Access Signalling System No 2* (DASS2) was developed by British Telecom to enable digital PABXs to intercommunicate with the local SPC digital exchange, thereby providing a common signalling system for digital PABXs across to a public PSTN. Its design allows maximum commonality to exist with the digital PABX to digital PABX signalling system, DPNSS.

DPNSS

The *Digital Private Network Signalling System* (DPNSS) is a common channel signalling system for digital PABXs, developed to enable such devices to intercommunicate across private leased 2 Mbits point-to-point circuits within the UK. DPNSS is based on a three-layer OSI architecture, the signalling information being transported between PABXs in time slot 16 of the 32-time slot channel circuit. This signalling feature can be used to convey a wide range of supplementary services in addition to the basic telephone and data services normally available from inter-PABX operation. Sophisticated error checking techniques within DPNSS ensure the signalling information's validity.

DPNSS was originally conceived by a consortium of UK digital PABX manufacturers and suppliers, including British Telecom, in advance of the CCITT I-series recommendations on private digital networks. DPNSS provides a sophisticated range of facilities for telephony and circuit switched data networks in a form which is compatible with ISDN standards, as well as following the specifications of the ISO reference model for Open Systems Interconnection. Its use of 2 Mbits digital circuits allows the PABX to allocate 30 digital voice channels with separate time slot channels for synchronisation and signalling. This approach is an innovation over previous inter-PABX signalling techniques, many of which combined the voice and signalling information using a method known as 'in-band' signalling, or alternatively utilised two separate circuits, one for speech and one for signalling.

Because of the commonality between DPNSS and DASS2, it is

intended to allow remotely located private networks to intercommunicate using DPNSS signalling conveyed over the interconnecting public network by DASS2.

Voice Transmission

To understand how voice transmission is carried across a digital communications link, let us first consider how voice communication was achieved by the older analogue method.

When a person speaks, the sound produced varies both in frequency and amplitude. The actual waveform of even the simplest word is a complex addition of a number of signals. In an analogue-based telephone system, the microphone in the mouthpiece of the telephone converts the sound waves into an electrical signal, which varies in almost exactly the same way as the sound signal. For reasons of economy, the existing analogue telephone network restricts the transmission of speech to a band of frequencies between 40 and 3400 Hertz. The higher and lower frequencies of the human voice are thus lost. This is one of the factors causing the typical telephone network distortion. Others include distance, electrical interference and general background noise.

The further an analogue signal travels, the more it is *attenuated* or reduced in size, until finally it disappears altogether. To overcome this problem, amplifiers or repeaters are placed along a transmission path to periodically restore the signal. The problem of noise associated with analogue transmission is more difficult to alleviate. Random noise can be introduced to every electrical signal by electrical disturbances such as office lighting, radio transmission and electrical equipment. To compound this problem, the amplification process for signals over long distances also amplifies the noise levels as a secondary effect. This could make the voice signal inaudible.

The digital communication method for voice transmission has several advantages over its analogue counterpart. Although the digital transmission of speech starts with the same microphone and conversion process described previously, the sound waves are not converted into electrical signals. Instead they are converted into a digital binary code representation of the sound wave. The conversion between voice and code is achieved by sampling the speaker's voice signal at regular intervals. The amplitude measured at each sample is then given a binary-coded representation. Assuming the use of the 'Pulse Code Modulation' (PCM) technique, which is still one of the most established techniques within PABXs, the sample of the incoming voice signal may be taken every 125 milliseconds, equivalent to 8000 times a second. Each sample is normally represented as an 8-bit binary word, which gives a total possible combination of 256 'quantum levels'. This binary code is then transmitted to the receiving PABX where the process is reversed, decoding the bit pattern to provide voice output. The device which undertakes this coding/decoding function is known as a *codec*.

Digital voice encoding has several advantages over the traditional analogue method. For example, the digital signal is less susceptible to noise because it is merely an encoded string of binary '0's and '1's, rather than an electrical waveform. Unlike the analogue signal, which needs repeaters over long distances, the digital signal can travel greater distances before it needs to be regenerated. At these regeneration points, the signal is recreated rather than amplified. This means that no distortion is carried forward.

The Codec

To provide an interface between analogue communication and the emerging digital technology, a device was required that would undertake conversion of either medium in either direction at potentially very high speeds. This device was known as a codec (coder/decoder) and one of its first applications was with the PTTs, providing a conversion interface between the analogue Frequency Division Multiplexing format and digital Time Division Multiplexing PCM techniques. Within the PTT networks these devices, termed Supergroup and Hypergroup codecs, provide an encoding and decoding function for circuits from 8 Mbits up to 140 Mbits, enabling a hybrid network of digital and analogue circuits to be created. In an end-user environment, the codec made its appearance in the PABX, providing the interface between the analogue channels from the telephone handsets and the digital circuits to the outside world.

With the evolution of the digital PABX, the codec is transferring from the PABX itself to a microchip within the telephone handset, providing conversion of the human voice at source. This enables a complete digital channel to be provided end-to-end across the network between any two suitably equipped handsets. This simple evolution has led to a new generation of digital PABXs being developed, which take advantage of the end-to-end digital channel as well as the introduction of the ISDN. The generic title for this device is the *Integrated Services Private Branch Exchange* (ISPBX).

USER ISSUES REGARDING DIGITAL PABXs

The ISPBX

The Integrated Services Private Branch Exchange (ISPBX) enables fully digital end-to-end links to be established, either by using the existing digital circuits of a private corporate digital network, or by employing the emerging public ISDNs. The ISPBX can, of course, still connect and communicate with the existing analogue PSTN and attached PABXs, but the ISPBX will then only provide the basic services.

The ISPBX can be installed either as a centralised or distributed exchange system, depending on requirements. With a centralised ISPBX, several thousand extensions could be installed locally to the exchange, with several hundred trunks or circuits linked to the local PTT exchange. These circuits could be either digital or analogue, depending on the PTT equipment. Assuming the local PTT exchange was not yet equipped for public ISDN services, a private ISDN and its associated facilities could still be employed locally around the organisation's ISPBX. In addition, using digital leased circuits, the private ISDN service could be extended to interconnect with other private ISDN services either internal or external to the organisation (see Figure 9.1).

The distributed ISPBX uses basically the same concept, locating smaller exchanges at remote geographical locations, and interconnecting these with 2 Mbits digital circuits. Control and supervision of the distributed system can be organised either at the local sites or via a central point. The distributed ISPBXs communicate over the private network using the DPNSS signalling system. Where a distributed ISPBX system is a requirement, it provides the network designer with an additional economic justification for the use of 2 Mbits digital circuits. Using ISPBX, such a system provides the integration of voice, data, text and image that the organisation may require from the network, thereby enabling many new distributed office automation services to be introduced as an overlay to the network services. Where public ISDN services are available, ISPBXs can be connected directly into the service, using the DASS2 signalling technique.

THE DIGITAL PABX

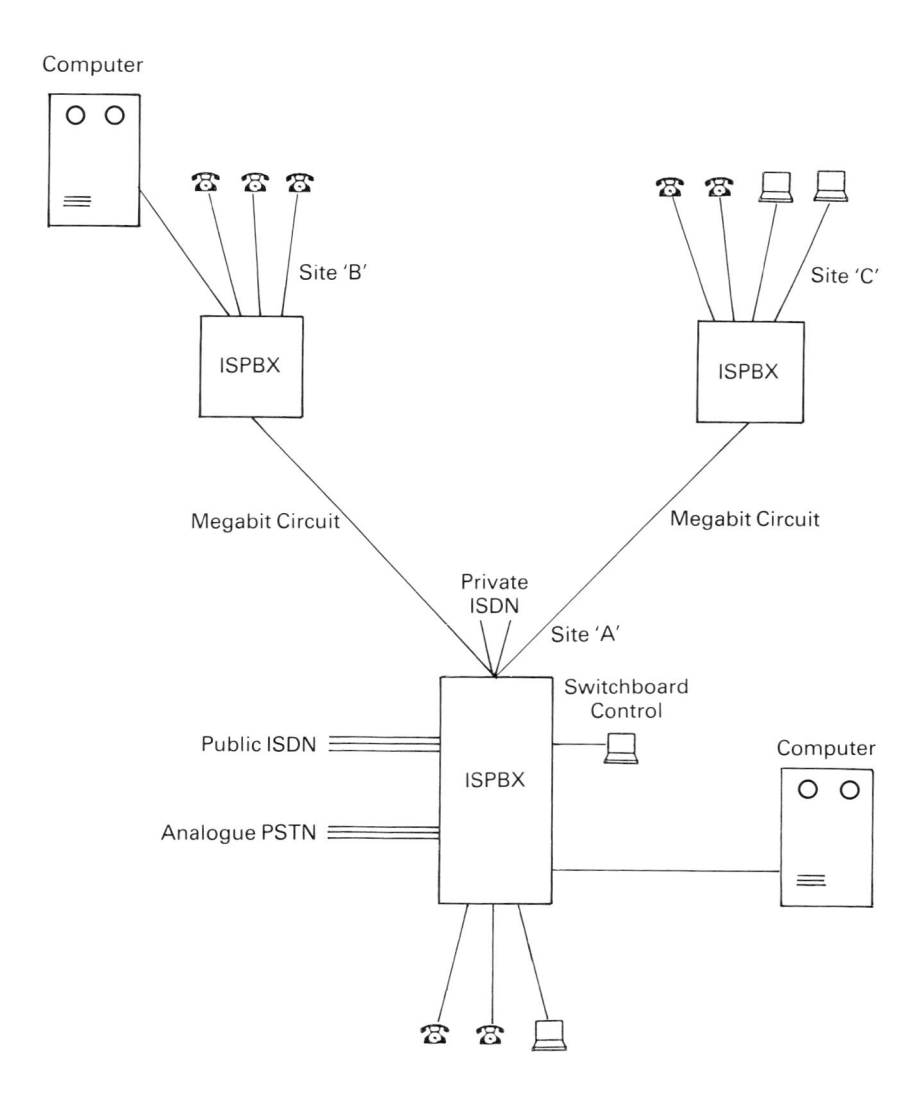

Figure 9.1 Example of a Private ISDN Using Interlinked ISPBXs to Provide Multiple User Services

Integrated Services Featurephones

With the introduction of the ISPBX, a new range of desktop telephones were introduced to utilise the ISDN facilities. This range of equipment is commonly termed the *Featurephone*. Although the Featurephone differs in functionality within the various product ranges from each manufacturer, it generally comprises a handset, a keypad for dedicated and programmable functions, and possibly one or more interface ports, usually either V.24 or X.21 compatible.

The voice signals entering the Featurephone are digitised within the device itself, enabling a digital route to be established through to the local ISPBX. The signal travels over a single pair of existing wires and can be simultaneously transmitted alongside data traffic. Transmission speeds on these two channels can be up to 64 Kbits per second.

In addition to the many functions available on standard phone sets connected to PABXs today, the Featurephone in conjunction with the ISPBX may have options such as incoming call detection, even when an outgoing call is being made; a LED display panel to show information such as number dialled, time of call, and text messages from other subscribers; and single button access to many other facilities. All this functionality is still powered only by the normal line power operation, enabling the system to remain available even during mains power failure.

Combining PABXs and LANs

Although digital PABXs and Local Area Networks (LANs) have evolved from very different backgrounds and requirements, the functionality of each system is being increasingly extended to include the features of the other. For example, the digital PABX can now provide many of the data communications facilities normally associated with a LAN while utilising the existing telephone wiring installed within a particular building. This dramatically cuts the initial cabling costs when compared with the normal LAN cabling requirements. Although the performance of the PABX is not equivalent to many LANs, it does provide a solution for organisations that require occasional, fairly short, data calls between computers or other office automation equipment.

The PABX's ability to provide direct gateways to the outside world via digital circuits is another plus in the PABX's favour. The introduction of ISDN and Integrated Services PBXs will further enhance this facility by providing a 64 Kbits switchable link to any location on a public or private network. The developments in LAN technology include the ability to handle voice communications over a local network. This is achieved by using the digitised voice techniques originally developed for the wide area networks. LANs are also being provided with gateways to enable them to connect with the outside world. This may take the form of LAN-to-LAN intercommunication over leased digital circuits, or the use of private or public packet switching data networks, giving data communication on a switchable wide area basis. ISDN is also increasing, providing a switchable digital service to the local area network user.

Taken together, the developments in digital PABXs and LANs can be used to provide the best of both worlds. Hybrid PABX/LAN systems are already available; in these a digital PABX provides individual 64 Kbits channels over conventional wiring to individual users. In addition, other inputs to the PABX are directly connected LANs, providing voice and data communications to a group of users. The PABX acts as a common gateway to the LANs, linking them to each other as well as to other parts of the communications network as required. Ultimately, the merging of digital PABXs and LANs should provide an increasing ability to support switching facilities at higher and higher speeds, enabling such devices to support not only voice and data, but video communication as well.

10 The Drive towards ISDN

GENERAL BACKGROUND TO ISDN

Given that the Integrated Services Digital Network (ISDN) is just around the corner in the telecommunications world; that there is a large and ongoing capital investment programme in ISDN by the PTTs; and that similar amounts of capital and commitment are being placed in ISDN product development by the telecommunications manufacturers; what effect, if any, is ISDN going to have on the most important group in this new environment — the end user?

ISDN in its broadest sense means many things to many people. Its fundamental objectives are to provide a common means of access to the network via a limited number of multipurpose interfaces; to provide a standard switching, signalling and transmission system across the network; and to incorporate into this a wide range of services and facilities which will allow any type of message, be it voice, data, text or image, to be transmitted without any constraint being imposed by the network.

Progress towards this ultimate concept is being achieved by the PTTs in an evolutionary rather than revolutionary manner. This approach means that the transition period to a full ISDN service is extended over many years, even decades. At the end of this time, however, the changeover should have been relatively smooth and cost effective both to the PTT and to the subscribers.

To gain the benefits of a smooth and cost effective transition, the ISDN must be able to work together with the existing networks to ensure continuity of service to the subscriber. From the PTT's viewpoint, this also enables a phased introduction plan to be drawn up, allowing equipment and plant from existing networks to be switched to the ISDN as and when required. But even though the introduction of ISDN is to be done in a progressive manner, the resulting effects could indeed be classed as revolutionary.

To date, most telecommunications networks throughout the world have been dominated by the telephone service. The many and various non-voice services, although showing considerable growth over the last few years, still remain a minority of the overall trafffic volume. ISDN could nevertheless be said to favour non-voice services, for although it provides many additional benefits for voice communication systems, it has an even greater impact. This is because it has a converging effect on telecommunications, office automation, data processing and new emerging services such as real-time interactive facilities, visual conferencing facilities and a host of other developments, all of which make use of the single ISDN link. Its general purpose utility combined with capacity and flexibility suggest that ISDN's future is geared to the development of non-voice services, whilst still catering for the traditional voice requirements.

To understand how the objective of ISDN is being put into practice, let us first examine how the early digital technology evolved to the state where the ISDN concept could be first put forward. We will then see how the move from concepts to the ISDN utopia can be broken down into four distinct stages.

In the Beginning . . .

The majority of the PTTs throughout the world are currently at various stages of the transition from analogue to digital transmisson. The process started back in the early 1960s with the introduction of *Pulse Code Modulation* (PCM) systems for short-haul, inter-exchange connections. These early PCM systems were initially used to increase the capacity of existing cables on short distance routes. However, it soon became apparent that they could provide the basis for an integrated network using digital transmission, switching and signalling on a national and eventually worldwide scale.

In North America, this technology was quickly implemented by AT & T, basing their design on 'Mu-law' encoded 24 channel systems operating at 1.544 Mbits. In the UK the manufacturers of transmission equipment (in collaboration with the Post Office) followed the trend set by AT & T.

THE DRIVE TOWARDS ISDN 135

They adopted a 24-channel system, but based instead on 'A-law' encoding techniques. This diversion in encoding methods at such an early stage resulted in the development of the two separate digital hierarchies we see today. The 'Mu-law' system is dominant in North America and Japan. The A-law system, now based on a 30- channel design operating at 2.048 Mbits, is standard throughout Europe, Australia, South East Asia and other areas. The result is that there are two encoding methods which cannot interwork directly because of significant differences in their design characteristics. In the international standards arena, CCITT was obliged to adopt both as international recommendations, as well as defining interfaces between the two.

The technology of digital switching was developing in parallel with the evolving digital transmission technology. This technology was based on solid-state crosspoints being installed as replacements for the old mechanical reed-relay cross-bar switching systems. The main driving force behind this evolution was the potential cost saving involved in the introduction of digital switching technology. This in turn also heralded the development of digital signalling methods, particularly those developed under the sponsorship of the international standards organisations.

Stage 1: IDN

From the growth in digital transmission and the parallel development in digital switching and common channel signalling, there evolved the concept of a multi-purpose *Integrated Digital Network* (IDN). In an IDN, switching nodes are linked by high-speed digital transmission circuits to local processor-controlled digital exchanges, thereby providing a PTT with a physical backbone on which to build emerging digital services. The British Telecom version of IDN is based on System X and System Y digital *stored program control* (SPC) exchanges, which together with the cross-connection sites and high-capacity digital trucks provide a fully digital network between these types of exchange.

Subscriber access to the IDN was initially via the existing analogue circuits from the subscriber premises to the local SPC exchange. The actual subscriber benefits at this time included improvements in circuit quality, call set-up times, and flexibility and access to some additional supplementary services.

Also within the BT IDN structure is the *National Digital Private Circuit*

Network (NDPCN), a synchronous digital network which is used as the basis for the KiloStream and MegaStream services. Although structurally part of IDN, the NDPCN is functionally separate and provides the subscriber with leased point-to-point digital circuits that will continue to form the major part of the BT digital private network, at least in the near future. The NDPCN consists of a number of cross-connect sites in central BT exchanges throughout the UK. From these sites, NDPCN digital circuits radiate out to additional primary multiplexer sites. This allows for the more or less total coverage of the UK. The fact that NDPCN is part of IDN does, however, allow NDPCN services to be connected into the developing ISDN as and when required.

Stage 2: IDA

ISDN is the result of the work undertaken by the International standards bodies such as the International Telegraph and Telephone Consultative Committee (CCITT), the International Radio Consultative Committee (CCIR) and the European Conference of Posts and Telecommunications (CEPT), to name but a few. However, there is one major drawback with standards-making: a great deal of time is taken to formulate and refine an idea for an ISDN requirement into an international recommendation, and even more to develop that recommendation into an ISDN product or service. The whole process is a fairly long and arduous task and in many cases it can (and does) take years.

As an example of this, the initial set of ISDN recommendations became available at the end of the 1980–1984 CCITT study period, when they were ratified by the CCITT Plenary Assembly. These were called I-series recommendations and published in the CCITT Red Book. The problem here was that until a complete set of stable recommendations covering a wide range of ISDN-related subjects ranging from user-network interfaces to maintenance issues was available and internationally agreed upon, no organisation had the ability to actually install a fully functional implementation of ISDN. As an interim solution, a number of PTTs decided to undertake field trials of ISDN-type networks, basing the design on existing and emerging standards but also incorporating proprietary solutions where necessary. One of the principal aims of such trials was for the PTT, and indeed the subscribers, to gain practical experience of what at that time was still just a concept.

From the mid-1980s onwards a number of pilot services were initiated,

THE DRIVE TOWARDS ISDN

among them British Telecom's Integrated Digital Access (IDA), a method of supplying subscribers with a digital access route to the IDN, but incorporating with this a number of ISDN-type services. IDA took the form of extending the digital link from the local British Telecom SPC exchange to the subscriber's premises, thus providing two end users with a fully digital 'end-to-end' link, either switched or non-switched, operating at channel speeds of up to 64 Kbits (see Figure 10.1). To this basic interconnection are added the services and gateways needed to provide a usable network service. One of the main features of IDA was that it could still provide subscribers with access to existing BT networks such as the PSTN or Packet SwitchStream. This allowed IDA subscribers to connect their telephony, data, text and low-speed image devices via their IDA access point, and in theory removed the need to retain their existing circuits for this equipment.

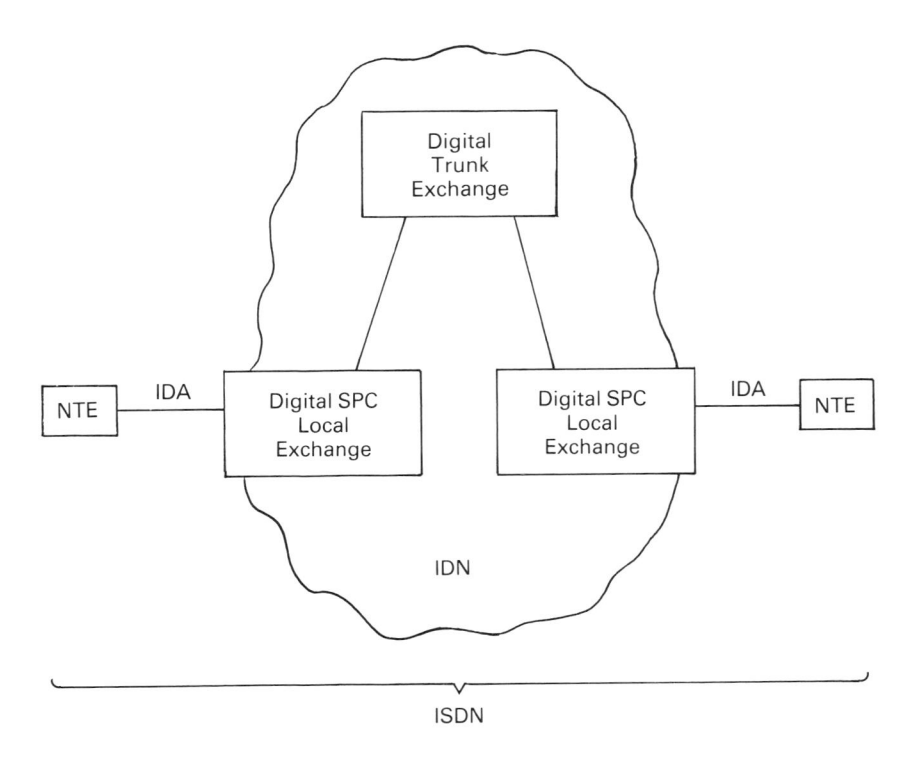

Figure 10.1 The Relationship between IDN, IDA and ISDN

Stage 3: ISDN

The eventual aim of the international development of ISDN was to provide a single circuit connection operating at 144 Kbits, which in turn would provide the subscriber with two 64 Kbits channels together with a l6 Kbits signalling channel.

In actual use, this single connection provides two high-speed digital circuits which enable two simultaneous data connections or one voice and one data connection to be made to the same or two separate destinations. The data traffic can take the form of data, text or image (eg photo videotex or high-speed facsimile). The two high-speed channels are commonly known as the *B-Channels*. The primary function of the l6 Kbits channel is to provide a signalling path for the B-Channels between the subscriber's equipment and the local PTT exchange. This pathway enables the ISDN to exchange the control and monitoring information necessary to manage the system.

A secondary function of this 16 Kbits channel, known as the *D-Channel,* is to utilise its spare transmission capacity by providing low-speed services such as packet data transmission, operating at speeds of up to 9.6 Kbits. This low-speed data service could be routed to either of the B-Channel's current destinations, or to a completely separate destination anywhere on the network.

Stage 4: Broadband ISDN

In conjunction with the introduction of ISDN, a future evolution strategy for ISDN is being developed which will involve the integration of switched broadband services into the integrated services network.

Broadband services will require bit rates of between 1.5 Mbits and 140 Mbits, as well as expansion of the transmission media over long distance, local and subscribers' lines. The transmission medium best suited for this task is optical fibre cable, which is one of the reasons why British Telecom (along with many other PTTs throughout the world) is rapidly installing such trunking. The services offered in broadband ISDN will include all the 'narrowband' services outlined for ISDN today. In addition, there will be a large number of new services which can be broadly divided into *dialled* broadband services for individual communication and *distributed* broadband services for mass communication.

THE DRIVE TOWARDS ISDN

The most relevant of these two groups for the commercial and financial organisation is the dialled broadband services. These may well include video telephony in a practical and cost effective form. It will have all the features available from standard telephone sets, plus a high quality video and voice connection, all built into a multi-function workstation. Video conferencing would be an extension of video telephony, allowing a number of users to interconnect voice and video links. A split-screen technique would be used to display the various participants. An expansion of studio video conferencing would also be available, allowing switched connections and multi-locations to be used. Other possible uses could include broadband interactive videotex, ultra high-speed transmission on a switched point basis, and distributed broadband services such as high-definition TV programme distribution and stereo programme broadcasting.

To achieve the integration of broadband and narrowband ISDN, a number of possibilities have been proposed:

— A separate broadband network which would be installed and expanded as required, independent of other networks. This has the disadvantage of totally neglecting the integration aspect.

— A solution based on the West German Bundespost BIGFON trial system, where the broadband and narrowband links are integrated on the user's line, but the networks are separate within the PTT's internal network.

— To totally integrate the broadband network into ISDN, supplementing the customer's '2B+D' link with a '30B+D' broadband circuit. This would have a number of technical and cost implications for both hardware and software within ISDN.

Whatever the final structure, broadband communication will be the next development step after the integration of the narrowband services in the ISDN. Its technical feasibility has already been demonstrated by a number of pilot projects; the technology for a broadband network is already available or under way (European involvement in this area includes a broadband peripheral switching project as part of research initiative 'Eureaka' and projects under 'Race', the European telecoms research programme); and the various standardisation bodies will produce recommendations within a few years.

TECHNICAL CONCEPTS BEHIND ISDN

IDA

In the initial IDA specification, the subscriber had the option of either single-line or multi-line IDA connection. The single-line (Basic Rate Access) IDA utilises a single pair of existing local telephone cables to carry an 80 Kbits full-duplex link. This circuit comprises one high-speed 64 Kbits channel suitable for voice or data (the B-Channel); one low-speed channel of 8 Kbits for data only (the B-Channel); and an 8 Kbits signalling channel (the D-Channel). The single-line IDA is supplied with an NTU equivalent known as the *Network Termination Equipment* (NTE). This is sited on the subscriber's premises and provides standard CCITT X and V-series interfaces for connecting subscriber terminal equipment. The NTE provides termination of the incoming digital circuit from the local exchange, termination of subscribers' equipment, and conversion of subscriber equipment protocols and data rates to protocols and rates as required by the network.

Initially two NTEs were developed, designed to support both existing terminals and new terminals which were optimised towards ISDN. The first of these, NTE1, consisted of a digital telephone, keypad, display and a single data port which could be configured to support a variety of terminals to CCITT X.21 bis and X.21 (leased line variant) interface standards. The NTE1 was designed during the early stages of the ISDN development programme, and was not intended to be used beyond the pilot service stage.

The NTE3, unlike the NTE1, had no telephone, keypad or display. Instead it provided six ports, each capable of supporting terminals conforming to one of the CCITT recommendations X.21 bis, V.24/V.28, X.21 (leased line variant), X.21 (circuit switched variant) or 2-wire analogue standards. Because single-line IDA supports only two data circuits, no more than two of the six terminal ports could be connected to the exchange at any one time, the connection control being resolved by the NTE, which acts as a traffic concentrator. Terminals without call set-up capability required an intermediate device called a remote selection unit.

The next NTE was NTE4, which has two X.21 data ports; initial versions are provided with adaptors to support other interfaces. Control signalling is carried within the data channel to the interface, then

THE DRIVE TOWARDS ISDN

separated within the NTE onto the signalling channel out to the local exchange.

I-series terminals will be supported by new designs of NTE as and when the need arises. This includes NTEs capable of supporting a number of terminal devices with a universal access interface for voice, data, text and image, for use on public and private networks.

The other version of IDA is known as multi-line IDA, and was intended mainly for organisations that require data and telephone services and have a PABX to distribute the calls. The multi-line IDA (Primary Rate Access) services terminate on a new generation of digital PABX generally known as an Integrated Services Private Branch Exchange (ISPBX), which provides the capability to route voice, data, text or image to any of its extensions. Each multi-line IDA provides the subscriber with thirty 64 Kbits channels, these channels being handled by the use of time division techniques over a 2 Mbits PCM link. Thus time slots 1 to 15 and 17 to 31 support the thirty user channels, whilst time slots 0 and 16 provide for the two remaining signalling, synchronisation and monitoring channels. The multi-line IDA connection is terminated at the subscriber premises by a network terminating unit (NTU), which constitutes the NTE for multi-line IDA, and presents a standard CCITT 2 Mbits interface to the subscriber.

British Telecom has adopted international CCITT recommendations G.703 and G.734 for the 2 Mbits interface and transmission standards of multi-line IDA, but was working in advance of international signalling standards for interworking between the subscriber's ISPBX and the local SPC exchange. BT therefore developed its own signalling system, known as Digital Access Signalling System No 2 (DASS2), as an interim solution between the pilot IDA service and the availability of stable international standards.

DASS2

The 'Digital Access Signalling System' (DASS), as it was originally known, was developed to enable ISPBXs to intercommunicate with local BT SPC exchanges, thus providing a common signalling system across a public digital PSTN. The initial version was DASS1, and was intended to provide a signalling system for both the single-line and multi-line IDA pilot service. In practice DASS1 was used for the pilot single-line service only, utilising the 8 Kbits signalling channel for this purpose.

DASS2 evolved from this original version, having incorporated into it principles from other signalling system developments and the ISO reference model for Open Systems Interconnection (OSI). It was DASS2 that was then used to provide multi-line IDA signalling utilising the 64 Kbits signalling channel for this purpose (ie time slot 16 of the 2 Mbits circuit). As well as providing signalling for ISPBXs, DASS2 also interconnects to NTE devices, including individual handsets, and is designed to have maximum commonality with the Digital Private Network Signalling System (DPNSS) developed for private network ISPBX to ISPBX signalling. This commonality factor is intended to enable DPNSS signalling to be conveyed over the public networks by DASS2, thus allowing separate remotely located private networks using DPNSS to be interlinked across the public network.

DPNSS

DPNSS is a common channel signalling system for ISPBXs, developed to enable such devices to intercommunicate across private digital networks in the UK. DPNSS was originally conceived by a consortium of UK digital PABX manufacturers and suppliers, including British Telecom, in advance of the CCITT I-series recommendations on private digital networks signalling systems. DPNSS provides a sophisticated range of facilities for telephony and circuit switched data networks in a form which is compatible with ISDN standards, as well as following the specifications of the OSI reference model. DPNSS makes use of the CCITT recommendation I.431 primary rate layer 1 standards to carry the D-channel layer 2 and layer 3 'UK proprietary' protocols.

ISDN

The ISDN specifications are contained in the CCITT I-series recommendations outlined in Figure 10.2. These are structured so that new and developing recommendations can easily be added to the appropriate sections. (In addition, Appendix 4 provides an overview of the I-series recommendations.) The 144 Kbits digital circuit, which is the subscriber's basic access to ISDN, is subdivided into two 64 Kbits voice or non-voice B-channels and a 16 Kbits D-channel for outband signalling and low-speed services. This '2B+D' channel structure forms the basis of the CCITT recommendation I.412, 'ISDN user-network interfaces'. To

THE DRIVE TOWARDS ISDN

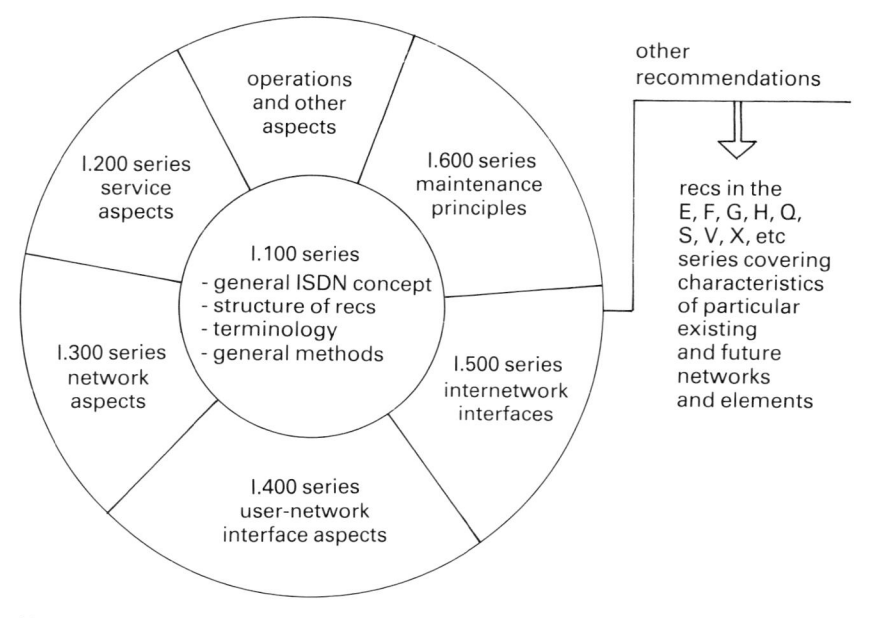

```
                    operations               other
                    and other                recommendations
                    aspects
  I.200 series                  I.600 series
    service                     maintenance      recs in the
    aspects                     principles       E, F, G, H, Q,
              I.100 series                       S, V, X, etc
              - general ISDN concept             series covering
              - structure of recs                characteristics
              - terminology                      of particular
  I.300 series - general methods                 existing
    network                     I.500 series     and future
    aspects                     internetwork     networks
                                interfaces       and elements

                    I.400 series
                    user-network
                    interface aspects
```

Note:
- models
- reference configurations
- tools, methods
  are contained in the appropriate I - series recommendations

Figure 10.2 Structure of I-series Recommendations

meet the wide-ranging bandwidth requirements for subscriber access to ISDN, the CCITT has standardised on a family of user-to-network interface structures. Of these, the basic interface structure (I.420) and the primary interface structure (I.421) are likely to be of most interest to users. To assist definition of ISDN user-network interfaces, CCITT produced recommendation I.411, a reference model for basic access interface structure. This model is outlined in Figure 10.3.

CCITT recommendation I.420 defines the interface for basic access to ISDN. It permits direct terminal access to the 2B+D channel structure, offering separation of control signalling from the data channel and a bus

```
                    S           T
                    |           |
                    |           |
  ┌─────┐           |  ┌─────┐  |  ┌─────┐
  │ TE1 │───────────┼──│ NT2 │──┼──│ NT1 │──── Digital
  └─────┘           |  └─────┘  |  └─────┘     Circuit
                    |     |     |
              R     |     |     |
              |     |     |     |
              |     |     |     |
  ┌─────┐     |  ┌──┴─┐   |     |
  │ TE2 │─────┼──│ TA │───┤     |
  └─────┘     |  └────┘   |     |
```

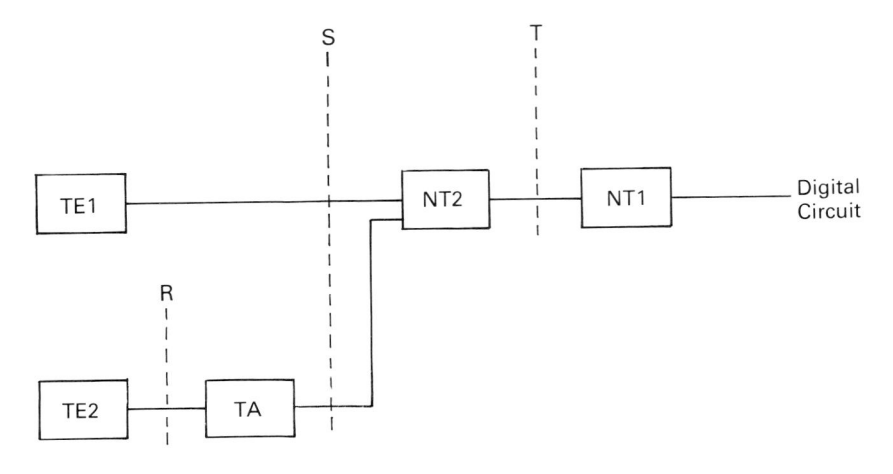

Notes:

TE1	— terminal equipment supporting I-series ISDN interface
TE2	— terminal equipment supporting X or V-series interface
TA	— terminal adaptor for interfacing X or V-series to I-series interfaces
NT2	— distribution network, possibly an ISPBX, LAN or passive bus structure allowing point-to-point or point-to-multipoint communication
NT1	— NTE which handles line transmission termination, maintenance functions, timing etc
'R, S, T'	— reference points where customer network interfaces may occur. 'S' and 'T' are I-series interfaces whereas 'R' may be non I-series CCITT interfaces eg X.21, X.21 bis etc

Figure 10.3   CCITT Model for 144 Kbits Basic Access to ISDN

configuration enabling up to 8 terminals to share one network access point.

CCITT recommendation I.421 defines the primary rate access from ISPBXs and other similar complex devices at 2 Mbits. I.421 uses the same signalling techniques as the basic I.420 access, except that the signalling channel rate here is 64 Kbits rather than 16 Kbits. The electrical characteristics of the primary rate interface are specified in accordance with the G.703 recommendation for European thirty-channel PCM systems. (Note that in North America and Japan the primary rate access will follow convention at 1.544 Mbits.)

Although I.420 and I.421 vary widely in their electrical and physical characteristics, both structures share the same D-channel link access protocol and signalling procedures.

International Signalling Standards

International standards will eventually be available as an alternative to the UK 'de facto' standard of DPNSS. These international standards will reflect the OSI reference model, particularly the physical layer (layer 1), the data link layer (layer 2) and the Call Control Procedures (layer 3).

Layer 1 will be based on CCITT I.431, the same standard as currently used by DPNSS. The main differences between DPNSS and the international inter-ISPBX signalling system will be at layers 2 and 3 (signalling protocols). Layer 2 will conform to CCITT I.441 while layer 3 will be based on CCITT I.451. The problem at layer 3 is that regional variations are likely due to differences in the requirements of various PTTs. Supplementay services at this layer will be introduced as required.

The eventual switch from DPNSS to international standards in the UK will largely depend on end-user requirements. The initial demand is seen as likely to come from organisations with international private circuits. Even then it would be possible to run both DPNSS and the international equivalent on ISPBXs if required, thus negating some of the pressure to adopt international standards. For British Telecom's ISDN network, the replacement of DASS2 with the CCITT Signalling System No 7 would depend on internal BT policy decisions at that time.

USER ISSUES REGARDING ISDN

The IDA Experience

The IDA service from British Telecom gave UK organisations their first experience of a public switched digital network. Although not a complete service when compared with the full ISDN standards, it provides valuable lessons for both the users and the provider. From the users' point of view, the following section outlines the lessons learned from these initial experiences in voice, data, text and image services over IDA.

Voice Services

The benefits of the IDA voice service are best exemplified by comparing it with the traditional PSTN service. The quality of digital transmission, with its lack of background noise and clarity of speech, is far superior to that generally obtained by analogue methods. Call set-up times are dramatically improved. Traditional analogue set-up times of between 15 and 30 seconds are reduced to around one second when using a combination of IDA and digital telephone handsets.

The full ISDN voice service also supports a number of supplementary services. These include features such as 'call waiting with caller number identification', where a caller to a busy telephone can notify the called party of his presence by means of an audible or visual signal on the called party's handset, in addition to a visual display of the calling party's number. If the feature 'registration of incoming calls' is activated, then when the called party's telephone is busy or unattended the associated exchange logs any calling party's number together with the date and time, for later access by the called user. 'Call forwarding' to another telephone on the ISDN, and 'call back when free' where the exchange automatically sets up a connection between the party that called earlier and the busy telephone after the latter has become free, are other supplementary voice services. All of these are possible because of the functionality supplied by the signalling D-channel. Such supplementary services may, however, be chargeable extras, the tariff charges being dictated by the PTT.

The negative aspect to all these benefits is the fact that cost savings (if any) obtained by an organisation using IDA voice services in preference to traditional PSTN voice services are generally unquantifiable. Perhaps the best equation to use in this case is 'time saved against IDA cost'.

Data Services

For data calls over IDA, the benefits are more quantifiable. The IDA ability to provide a switchable 64 Kbits channel gives the end user the combined advantages of traditional PSTN call flexibility together with BT's point-to-point high-speed digital KiloStream service. High-speed file transfer between computers can be obtained on a dial-up basis, thus removing the requirement of some organisations to have permanent leased circuits between remotely located devices. IDA can also provide a dial back-up service to organisations that wish to retain leased circuits such as KiloStream, but that also require a 64 Kbits back-up alternative in case of leased circuit failure.

Early versions of IDA did suffer from some limitations, such as the need for the calling and called parties to verbally discuss the procedure to set up a data call before both parties switched over from voice to the data call itself. However, new releases of IDA would rectify these problems.

Text Services

Text can be transmitted as 'unstructured data' or as an image. Although the data form is a faster method of transmission, image in the form of Group 4 facsimile does provide excellent results. The Group 4 facsimile machines can transmit at both 8 Kbits and 64 Kbits, thus giving high and low definition output depending on the quality required. At the higher speed a standard A4 page can be transmitted in around three seconds (compared with 30 seconds on a Group 3 machine).

Documents to be transmitted are scanned, compressed digitally, and then stored on an inbuilt hard disk before the machine initiates a call across IDA. Received documents are similarly stored before being reverse processed and printed. As the documents are held on disk, they can be retrieved or retransmitted as required. Some machines have a compatibility option to allow interworking with Group 3 facsimile machines. Videotext compatibility is also available if required. The major disadvantage of the early Group 4 facsimile machines is their cost; they are significantly more expensive than their Group 3 counterparts. The Group 4 machines are therefore generally used only on applications where very high quality definition is required.

Photo Videotex or 'Picture Prestel' is a great improvement over the original videotext concept. The slow transmission speed and line quality problems sometimes encountered in the early systems meant that navigating through the menu hierarchy was often a slow and laborious task. The high speed and quality transmission of IDA greatly improves the videotext concept (a full videotext frame is transmitted in around five seconds), and added to this is the ability to incorporate high resolution 'still' pictures together with text in a videotext frame. The still pictures can be in the form of colour photographs that have been digitised and placed on to a photographic library held on a central database. Computer-designed high-resolution graphs and diagrams can be stored in a similar manner. Such technology opens up new avenues in the ability to store and forward such information within organisations. It could also form the basis for new public services.

The full ISDN teletex is an enhancement of the existing teletex service, which in turn was a step above the original telex. The ISDN teletex can transmit an A4 page at 64 Kbits in around one to two seconds. Its protocols, coding and resolution are based on the CCITT recommendations for Group 4 facsimile. The full ISDN textfax is a combination of text and facsimile communication. Its main application would be for transferring documents which contain text and diagrams, and possibly handwritten notes and drawings as well. It is presumed that textfax will become the tool for speech and simultaneous document transmission in an office environment.

Image Services

With regard to the image applications of IDA, slow-scan TV used in the area of security and surveillance is probably one of its best applied uses. A single security office can monitor a number of remotely located cameras, each with a link into the IDA. Calls across the network can be initiated at either end, for instance by the security officer as part of a routine pattern around each camera site, or automatically by the camera end if an alarm is tripped. The IDA allows the security officer to send control commands (pan, tilt, zoom etc) to the remote camera as well as operating remote switches such as lights.

The IDA 'closed user group' option on originating and terminating access points ensures that external callers cannot simply 'busy' the line by making a call to it, thus making the security system itself secure. Slow-scan TV also has applications in the medical world, where live, detailed pictures can be sent between medical staff situated many miles apart. Slow-scan TV picture refresh rates are user selectable. The refresh rate ranges from once every 6 seconds down to once every .75 seconds, and the longer time periods give better picture definition.

Conclusions Drawn from IDA

One of the main design criteria of IDA was that it would give the subscriber access not only to any other IDA user, but also to any of the other existing networks. In practice, this meant that an IDA terminal should be able to connect through to BT networks such as PSS, and that an IDA telephone handset would interwork with the existing PSTN, admittedly with reduced functionality when such calls were made.

Negative aspects did emerge from the IDA trials. Some organisations felt constrained by the lack of facilities within the service, especially when

THE DRIVE TOWARDS ISDN

both a 64 Kbits data channel and a voice channel were required concurrently but only one of the requirements could be serviced. This problem was offset to some extent by the availability of the 8 Kbits data channel, a fairly high-speed service when compared with the speeds obtained from telex or a V.22 modem connection. The main concern for many IDA subscribers was upward compatibility with the full ISDN service. The IDA products had to be upgradable in order to ensure that the organisation concerned would be able to gain maximum utilisation from ISDN as it is introduced.

From the British Telecom standpoint, the IDA pilot service was seen as the initial ISDN service that would eventually evolve into the full 2B+D ISDN system. Existing digital networks such as PSS and KiloStream would conceptually form part of the ISDN. However, because of the economics involved, they would probably remain as separate networks for the foreseeable future. BT's Interstream Two (the packet network adapter), which was originally installed as a PSTN/PSS gateway for teletex services, now also provides this interworking between the ISDN and the PSS for both teletex and other services (Figure 10.4). 'Interstream Three' allows ISDN, PSS and PSTN teletex terminals to interwork with the telex network. In addition, 'Interstream One' provides interworking between the telex network and the PSS for asynchronous and X.25 terminals. This provides the ISDN with an alternative access route to the telex network via Interstream Three and One respectively.

Connection between the ISDN and point-to-point digital services such as KiloStream is provided by interfacing physical circuits within the NDPCN. This could provide private corporate digital networks with an extension and enhancement feature into the global ISDN. Interworking with the PSTN is an inherent feature. No special arrangements or

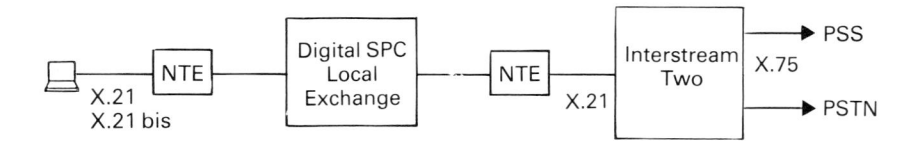

Figure 10.4 Interworking between IDA and PSS/PSTN Using 'Interstream Two'

interworking equipment are required. The ISDN numbering scheme forms part of the UK PSTN numbering scheme and no translation or prefix is necessary.

The Dawn of ISDN

The technology for ISDN is generally available now. Only the regulatory, standards and commercial issues still need to be completely resolved. Given this setting for ISDN, what are the tangible benefits now and in the immediate future to the end user? The following are some of the main considerations:

Tariffs: In an all-digital network where a bit is a bit irrespective of whether it represents a sample of speech or an element of a datastream, tariffs must move away from the service-dependent structure of the analogue environment to a tariff based on the volume of bits transmitted. As a significant increase in voice tariffs is politically difficult, the way should be clear to reduce non-voice tariffs to a standard level with voice within ISDN. This should reduce the overall usage cost to the business user.

Universal Connection: One of the fundamental objectives of ISDN is to limit the number of interfaces to the network, thus allowing the end user to connect a number of devices to the network via a standard ISDN socket. This gives the end user greater freedom in the choice and location of telecoms equipment.

Speed and Flexibility: ISDN offers the subscriber a high-quality, high-speed link which is user-switchable to any desired location on the network. It offers facilities superior to those provided by the current analogue-based PSTN, or by the high-speed, fixed, point-to-point digital services. ISDN circuits can also be utilised more efficiently than any existing dedicated circuit equivalent.

Improved Facilities: The 'D' channel's signalling ability enables a number of new facilities to be introduced to the subscriber. These include caller number identification and call forwarding on a national and eventually international scale.

Multiple Channels: The ISDN '2B+D' facility provides the subscriber with three channels, each of which could be routed to separate destinations. The standing rental charge for such a service is expected to be less than that for two analogue circuits.

Administration: As ISDN end-user devices become more multi-purpose, the natural tendency within organisations will be to converge separate communications departments into a single corporate group. Administration, management, procurement and service levels should therefore be simpler, more effective and ultimately cheaper.

A more natural way of working: IT in general and PCs in particular have tended to close people off from interworking with their colleagues and clients. ISDN offers a way to reverse this trend by allowing its combined services to provide a more natural interworking environment using its multiple communications facilities.

The Negative Aspects of ISDN

Some of the more negative aspects of ISDN should be weighed against the positive benefits. For instance, although ISDN could well go some way to solving many organisations' communication requirements, it may well, by its very nature, fall behind the current demands of the market. This is because of the timescales involved in the design, definition and implementation of features within ISDN. The elapsed time from the initial request to the standards bodies to the incorporation of an ISDN feature could well be measured in years. The conclusion is that for the time being the future of private corporate networks remains fairly secure, but they will tend to incorporate ISDN when and where necessary. Certain services which are either unavailable or too expensive from public ISDN suppliers could then be incorporated into a private network.

Another drawback can be seen at the applications services level of ISDN. Today these services tend to be provided either individually or in devices with a couple of applications services. Although integration exists at the physical transmission level, these voice, data, text and image services are not really drawn together to provide the end user with a single integrated ISDN applications service that is both simple to use yet powerful in its abilities. Considerable effort is still needed to bring about such integrated applications. Once available, and assuming reasonable pricing levels, the end user should then be able to obtain the full benefits from ISDN.

European Harmonisation

The European Commission has recognised the need for ISDN to be based on a European rather than a national scale. Working in conjunction with national administrations, the PTTs and industry an EEC recommendation has been prepared outlining a harmonised and phased approach to the introduction of ISDN in the Community. The European Commission has set the European PTTs an objective of 5% ISDN penetration in Europe by 1993, with a potential goal of around 90% of large and medium business users and 15% of residential users by the end of the century. However, it is customer demand which will dictate the pace at which ISDN is implemented, with the large business sector leading initial customer demand, and the residential sector following on. This could result in a 'chicken and egg' situation, with the PTTs unwilling to invest heavily and quickly in ISDN without customer demand, and customer demand being limited due to the restricted services within the public ISDN.

11 Digital Implementation

INSTALLING POINT-TO-POINT DIGITAL CIRCUITS

The introduction of point-to-point digital circuits into the corporate network may seem fairly straightforward at first sight, especially when the initial step towards digital corporate communications may be the installation of a couple of kilobit circuits between one or two company locations. Indeed, at this level, digital communication is a fairly simple matter of outlining requirements and needs, with the PTT installing and commissioning the link and the subscriber then attaching his equipment to the appropriate interfaces.

However, things are never as simple as they sound. It is important to consider how these one or two digital circuits may over time develop into a large digital network; how the digital communications will be controlled and managed by the organisation; and (one of the more important factors) how this new digital technology will integrate into an existing analogue communications network. (Unless an organisation is developing a digital network from scratch, a hybrid network is likely to exist for a significant period of time.)

Initial design and planning has always been considered the most important step in the successful implementation of any corporate communications network. With digital communications, this first step should be considered of crucial importance. The time and resources allocated to design and planning should in all cases be greater than those normally given to traditional analogue networks. Within this initial phase it is important to identify which aspects of digital technology are most applicable to the corporate communications requirements of the organisation in question. Once identified, the most appropriate techniques, equipment and systems can then be selected.

The negative aspects of digital communication should also be considered during this initial phase. It is important to identify these because in many cases they are very different from those normally encountered in analogue communications. This new set of communications problems has to be met with a new set of solutions. This chapter will cover some of the potential problems and pitfalls within digital communications design and planning, along with some possible solutions. Particular emphasis has been given to the area of analogue-to-digital conversion.

POINT-TO-POINT CIRCUITS

Single Channel Synchronous Data

Single channel synchronous data point-to-point analogue circuits operating at speeds of up to 19.2 Kbits have digital circuit equivalents which can be used as a direct 'one for one' replacement. This means that the additional advantages of digital transmission can be exploited with little capital outlay, other than the initial digital circuit installation charges. Depending on the PTT supplier, speed requirement and distance, a digital circuit may cost less to rent than its analogue equivalent, and as modems are not longer required additional savings can be made on both capital expenditure and maintenance cover. As a general illustration of cost savings, a British Telecom 9.6 Kbits KiloStream circuit has a rental charge breakeven point (compared with a conventional analogue circuit) at around 168 miles.

An enhancement option for this 'one for one' replacement is to upgrade the digital circuit speed, enabling a higher throughput to be achieved. However, one consequence of this option is the effect on DTE devices operating to CCITT V.24/V.28 specifications. As such, the specifications limit the maxium speed to below 20 Kbits, so the interface on each DTE device would have to be replaced by one of the higher-speed interfaces (such as X.21) if the full benefits of the higher-speed circuit were to be utilised.

A secondary consequence of increasing the circuit speed is the effect it may have on DTE performance. Take, for example, bulk file transfer between two computers. The throughput of data across the circuit will be improved by the higher speed, but the increased loading on the host processors may affect the performance of these devices. The increased

DIGITAL IMPLEMENTATION

traffic flow may also affect the overall performance of any attached FEPs, so it is important to consider upgrading DTE devices if and when upgrading digital circuit speeds.

An alternative to upgrading DTE interfaces is to attach basic kilobit multiplexers at each end of the digital circuit. The circuit itself either remains at its current speed or is upgraded to higher speeds of up to 64 Kbits. This approach eliminates the need to replace DTE interfaces, as the multiplexer provides the interface between the subscriber's equipment and the high-speed digital circuit. It also means that for a small cost (additional to that of the initial analogue to digital replacement) the circuit can now support several low-speed channels, or a combination of low-speed and voice channels, giving maximum utilisation from the circuit as well as providing potential for new services.

An additional consequence of upgrading to higher speed circuits may be the need to upgrade or replace older types of communications testing equipment. New equipment may be required that is capable of supporting the higher-speed interfaces. This type of upgrading may also apply to old patching or switching devices currently installed within the network.

The final consideration with all of these points is the need for staff training. The introduction of digital circuits, with their NTU/CTU devices, the possible use of digital multiplexers, switches etc, and the possibility of new or updated test procedures all mean that training must be provided to ensure that communications and operations personnel are competent with this new technology.

To summarise synchronous analogue-to-digital circuit replacement:

— replacement of circuits is a fairly straightforward and cost-effective operation;

— if circuit speeds are to be increased, the resulting 'knock-on' effects to DTE resources must be taken into account;

— basic digital multiplexers provide a cost-effective way of supplying additional channels between two locations;

— current communications test equipment and other network components may require upgrading or replacing;

— personnel training in digital technology is an important factor in successful digital implementation.

The majority of these points will be equally applicable to the following circuit configurations, and should be taken into account in any digital network design.

Single Channel Asynchronous Data

As digital circuits provide for synchronous transmission only, the replacement of single-channel asynchronous point-to-point analogue circuits can prove to be a more complex operation that might have been envisaged. As there is no direct digital counterpart to the asynchronous circuit, some form of conversion from asynchronous to synchronous must be employed. This is achieved either by installing a converter device between an asynchronous DTE device and the digital circuit, or, more commonly, by attaching multiplexers to each end of the circuit that have inbuilt conversion facilities for a number of attached asynchronous DTE devices.

Single Channel Voice

Single-channel analogue voice circuits connected on a permanent point-to-point basis are fairly unusual today. Most single analogue point-to-point circuits are used for data transmission, with an option for speech if required. Otherwise most corporate voice channels are routed along multiple circuits between the corporate PABXs. Where single-channel analogue voice circuits exist and a digital replacement is required, a fairly expensive conversion is needed to a high-speed 48 or 64 Kbits digital link, together with some form of conversion device at either end to allow speech transmission. With the costs involved, a more cost-effective approach would almost certainly be to connect both ends into the local PABX (if available) or alternatively to use the emerging ISDN.

Multi-channel Point-to-point Data

This mode of operation in analogue circuits provides for two or more conceptually discrete data transfer channels to be established across a single point-to-point circuit. There are commonly two methods of achieving multi-channel operation across analogue circuits: *secondary channel operation* and *modem-integral* multiplexing.

In an analogue secondary channel system, circuit bandwidth not required for the main data channel is used to provide a carrier path for one or more additional slow-speed data channels. Multiplexers are required to produce the equivalent service on a digital circuit. As digital multiplexers will be either time division or statistical (as opposed to the traditional analogue frequency division multiplexing method) it may be necessary to increase the data transmission rates of the secondary channels in some cases. The analogue modem-integral multiplexer does exactly as its name suggests, multiplexing several slow-speed channels onto one relatively high-speed analogue circuit. Here again the digital multiplexers will provide an almost identical service, although unlike the modem-integral multiplexer, most synchronous digital multiplexers are not 100% efficient.

MULTIPOINT CIRCUITS

Multipoint analogue data networks can be either 'star' or 'tree' in their topology structure. A network with one central location linked directly to all the remote points can be considered a 'star' network. Any configuration which consists of a central location connected to a number of remote points, which are then connected to a further set of remote points, can be called a 'tree' network. Such networks require an analogue mixing device at the central location, which allows the central modem signal to be passed down to modems at the remote points. There is no equivalent digital service currently available from the PTTs, but a similar system could be constructed using a digital mixing device if such a network was essential.

An alternative is to use point-to-point digital circuits, multiplexers, switches and similar devices in various combinations to provide a similar type of network. However, the addition of such communications hardware will dramatically increase the cost of the digital network, and this will probably defeat the reasoning behind the creation of the original analogue multipoint system, eg a network that was simple and cost effective.

In such a situation a new strategy is really required. Rather than replacing the analogue multipoint network with a digital 'lookalike', it is better to develop a corporate network which not only incorporates the multipoint terminals, but also other corporate voice and communications services. Integration is once again the keyword, and the resulting digital network will lead to a more cost-effective solution for the replacement of analogue multipoint data networks.

When considering analogue multipoint networks, it is often necessary to decide whether a single move to digital circuits is essential, or whether individual circuits may be converted over a period of time. This produces a hybrid network during the conversion period, but results in a more smooth and cost-effective changeover.

DATA CIRCUIT RESILIENCE

Data links established across analogue point-to-point leased lines have normally had a greater resilience to total communications failure because the attached modems are generally equipped with PSTN connection facilities. This enables alternative links to be established in 'dial backup' mode. Although transmission speeds obtained across the PSTN are generally slower than those achievable on leased circuits, it does provide some degree of resilience at a cost-effective price, as well as the additional option of analogue circuit switching to any suitably equipped location if required. Until digital circuit switching in the form of ISDN becomes widely available, the problem of providing alternative routeing for failed point-to-point digital circuits will remain. To provide resilience in this situation, the design of corporate digital networks becomes crucial.

One approach would be to provide a primary and secondary circuit between two locations. This would provide alternative routeing capability but at a very high cost in comparison with analogue dial backup. The approach normally favoured is to design digital networks with alternative routeing through other intermediate nodes in the network, thus providing resilience to most circuit failures. However, this option is still costly by comparison with analogue methods.

A concern with both alternative routeing methods is that the circuits from the user site may well pass through to the same local PTT exchange. If the exchange fails, all connected channels may be lost. Some organisations therefore ensure that alternative routeing circuits are connected to at least two different local PTT exchanges, or alternatively that a combination of circuits from different PTTs is used.

DATA AND VOICE SWITCHING

The ability to switch between data and voice on analogue circuits is normally provided by one of two methods, either by placing a switch at either end of the circuit to enable switching between the modem and a handset, or by using a modem with inbuilt data/voice switching and an

DIGITAL IMPLEMENTATION

attached handset. With both methods, either speech or data transmission must be selected; concurrent operation is not available.

When converting to digital circuits, the most cost-effective way of providing this facility is to use multiplexers with voice channel support. This would provide voice and data channels in concurrent operation. The only proviso is that the digital circuit must be of a higher rated speed (ie normally 48 or 64 Kbits). For point-to-point voice channels, and indeed the internal corporate voice network as a whole, the choice of voice encoding system (ie CVSD, PCM, ADPCM) and the bandwidth of the voice channels (ie 64 Kbits, 32 Kbits etc), are both down to the network designer. Only when the internal voice network is to be interconnected to a public PTT network does choice of encoding system and speed become important. The regulations regarding connection to public networks must be taken into account at the design stage. For instance, although calls received via the PSTN may be forwarded along private circuits to distant locations, it is illegal to make calls from distant locations along private circuits and then out onto the PSTN.

PUBLIC AND PRIVATE ISDN

The Integrated Services Digital Network (ISDN) basically falls into two camps, public and private. Public ISDNs are developed, managed and marketed by the PTTs, so the services offered and the tariffs charged will be decided by these organisations, influenced by the demand generated by their end users. Little demand will probably mean a slower development and implementation of services in the UK, whereas a high demand may force up prices in response to market forces. Either way, demand will only be generated if services are available and attractively priced. A 'chicken and egg' paradox may well occur, with limited demand by the subscriber being met with a reduction in investment by the PTT. Private ISDNs, on the other hand, are under the control of individual organisations. Assuming the basic infrastructure of digital circuits, the addition of suitable ISPBXs would enable ISDN services to be introduced. Interworking with the public ISDN can then be done as a phased process, or even on a selective basis, depending on corporate requirements.

MANAGEMENT INTEGRATION

Within many organisations, the corporate voice network and the corporate data communications network are regarded as two completely separate resources, controlled and managed by two unrelated departments. To achieve the real benefits offered by digital communications technology, integration must take place not only at the physical hardware and software level but also within the corporate departmental structure. The logical first step in any corporate communications strategy is therefore to combine the diverse departments into a single body, with a clearly defined line of management controlling a single group of communications technicians and administrators. Such changes may meet with resistance from some of the individuals or departments involved, but a single corporate communications department is vital if the maximum gains and savings are to be made from the introduction of digital technology. This department should have the wide range of skills needed to cover all aspects of corporate communications (ie not only voice and data, but telex, facsimile and other similar services). Both wide area and local area networks should be covered, as well as the central components such as PABXs, data PBXs and FEPs.

The integration of personnel should allow a more concise and thought-out corporate communications strategy to develop. It should take all aspects of this corporate resource into account, and enable a more effective operational group to be created. This group will be more responsive to user needs, more cost-effective in terms of equipment procurement, and more efficient in terms of fault response and general service levels. If the integration of the diverse communications departments within an organisation is not achieved in a controlled and managed form, advances in technology may very well force change. This is because integration in the form of ISDN and multi-function workstations will force individual communications departments into working together. The only way to prevent this would be to prevent the new digital technology from becoming established within the organisation, which ultimately would be detrimental to everyone concerned.

Appendix 1
Glossary

Accunet A set of digital communciation services available in the USA from AT & T.

ADPCM Adaptive Differential Pulse Code Modulation. A voice encoding technique generally used over private digital networks.

Aggregate speed The sum total of a number of low-speed channels.

ANSI The American National Standards Institute.

Application Commonly used in data processing to describe a job or task which can be performed by a computer or similar intelligent device.

Architecture Framework for a communications or computer system which defines its functions, interfaces and procedures.

Asynchronous transmission A data communications method where individual characters (each enclosed by a start bit indicator and one or more stop bit indicators) are transmitted at variable time intervals.

AT & T The American Telephone and Telegraph Company.

Attenuation	Decrease in magnitude of current, voltage or power of a signal (ie signal loss or fading).
BABT	British Approvals Board for Telecommunications.
Baseband	A transmission medium onto which information is digitally encoded. Only one information signal is present on the medium at any time.
Bit	Abbreviation of binary digit. Signal element of transmission in binary notation. Either '0' (OFF) or '1' (ON).
Bits per second	The transmission rate per second of bits over a physical medium (also termed bit rate, bit/s, bps).
Broadband	A transmission medium having a wide frequency bandwidth capable of carrying a number of simultaneous analogue channels on separate frequency bands (also termed wideband).
BT	British Telecom.
Buffer	A storage area used to accommodate a difference in the rate of flow of data between two transmitting devices.
Byte	A group of binary digits used to represent a character of information.
CCIR	International Radio Consultative Committee.
CCITT	International Telegraph and Telephone Consultative Committee. The data and telecommunications standards body of the International Telecommunications Union, which is an agency of the United Nations. Founded in 1956, it comprises at

GLOSSARY

	various levels the PTTs, private telecoms operating companies, telecoms manufacturers and international/regional telecoms organisations.
CEPT	Conference of European Postal and Telecommunications administrations. Founded in 1959, its aim was to establish closer relationships between European PTTs. It develops a common European stance for input to CCITT. It also provides for a European standard on CCITT recommendations where options exist. General aim to harmonise European services.
Channel	A logical information path capable of transferring data in one or both directions. Generally a number of channels are carried over a single physical circuit.
Channel speed	The rate at which information flows along the channel.
CODEC	Coder/Decoder.
CRC	Cyclic Redundancy Check. Method of checking data transmission validity and detecting transmission errors.
CTU	Circuit Termination Unit, Mercury equivalent to British Telecom's NTU, (see NTU for further details).
CVSD	Continuously Variable Delta Modulation. A voice encoding method generally available over private digital networks.
DASS	Digital Access Signalling System. DASS2 provides common channel signalling between a digital ISPBX and local PTT SPC exchange, via digital circuits.
DCE	Data Circuit Terminating Equipment.

DDS	Dataphone Digital Service. AT & T 56 Kbits point-to-point service in the USA.
DPNSS	Digital Private Network Signalling System. DPNSS provides common channel inter-PABX signalling via digital circuits.
DTE	Data Terminating Equipment.
ECMA	The European Computer Manufacturers Association.
FDM	Frequency Division Multiplexing. Analogue multiplexing technique.
HDB3	High Density Bipolar 3. A digital encoding format.
HDLC	High-level Data Link Control. ISO frame structure for OSI protocols.
High-speed circuit	The physical communications medium between two locations or devices.
ICBN	Integrated Communications Broadband Network. A generic reference to broadband ISDN.
IDA	Integrated Digital Access. British Telecom's version of ISDN.
IDN	Integrated Digital Network.
IPSS	International Packet SwitchStream.
ISDN	Integrated Services Digital Network.
ISO	International Standardisation Organisation.
Isochronous	Data transmission format where all signals are of equal duration and are sent in a continuous sequence.
Kilobit	Literal meaning one thousand binary digits. In communication terms, used to express speed of transmission.

GLOSSARY

KiloStream	British Telecom's Kbit digital point-to-point service.
LAN	Local Area Network.
Layer	ISO term meaning a set of logically related functions which are grouped together. Interfaces to and from the layer are standardised.
Link	Communications path between two locations.
Manchester encoding	A technique for sending information bit-serially, in which the data and clock signals are combined.
MCL	Mercury Communications Ltd.
Megabit	Literal meaning one million binary digits. In communication terms, used to express speed of transmission.
MegaStream	British Telecom's Mbit digital point-to-point service.
MTBF	Mean Time Between Failure.
Multiplexer	A device which divides a physical circuit into a number of logical channels.
NTU	Network Termination Unit. British Telecom's circuit termination and customer interface unit for KiloStream circuits. Conceptual equivalent to an analogue modem.
OSI	Open Systems Interconnection.
PABX	Private Automatic Branch Exchange. A sophisticated, processor-controlled, private telephone exchange.
Packet	A block of data with a defined format containing data and control information.

Packet switching	A technique of switching data within a network. Individual data blocks or 'packets' are routed through the network according to addressing information held within the packet.
PAD	Packet Assembler/Disassembler. A PAD permits user equipment without a suitable interface to connect to a packet switching network.
PBX	Private Branch Exchange. See PABX.
PCM	Pulse Code Modulation. A time division technique that allows analogue speech transmission to be carried in digital form over a high-speed circuit.
Point-to-point	Direct link between two points on a network (eg between a computer and a terminal).
Port	Part of data communications device which acts as an input/output connection.
Prestel	The British Telecom videotex service.
Protocol	A set of rules to ensure a meaningful co-operation between partners.
PSDN	Packet Switching Data Network. Generic name for this type of network.
PSE	Packet Switching Exchange. A device which performs packet switching on a network.
PSS	Packet SwitchStream. British Telecom's X.25-based packet switching data network.
PSTN	Public Switched Telephone Network. In the UK, primarily operated by BT.
PTT	Postal, Telegraph and Telephone Authority. Name given to telecommuni-

GLOSSARY

	cations authorities in Europe and elsewhere, which act as common carriers for telecommunications.
Pulse Code Modulation	See PCM.
Signalling System No 7	CCITT inter-exchange signalling protocol for use between PTT digital SPC exchanges.
SPC	Stored Program Control. Relates to a software-based PABX or local exchange.
STDM	Statistical Time Division Multiplexer. A multiplexer which dynamically assigns bandwidth to attached low-speed circuits which currently have information to transmit.
System X and Y	Digital SPC exchanges used by the PTTs.
TDM	Time Division Multiplexer. A multiplexer which operates by dividing the high-speed circuit's bandwidth into a number of time slots. The multiplexer then assigns a time slot to one of the attached low-speed circuits, which in turn is linked to a user device, eg data terminal.
WAL2	A digital encoding format.
WAN	Wide Area Network.
Wideband	See Broadband.

Appendix 2

Bibliography and References

British Telecom Engineering

Communicate, Communicate Publications

Communications, International Thomson Publishing

Communications Management, EMAP Business and Computer Publications

Data Processing, Butterworth & Co

Gandy M, *Choosing a Local Area Network,* NCC Publications, 1987

Gee K C E, *Local Area Networks,* NCC Publications, 1982

Handbook of Data Communications, NCC Publications, 1982

Hardy P, *Digital Private Circuits for Data Communications,* NCC Publications, 1985

ISDN — The Deutsche Bundespost's Response to the Telecommunications Requirements of Tomorrow, Deutsche Bundespost

Lane J E, *Corporate Communications Networks,* NCC Publications, 1984

Lane J E, *The Integrated Services Digital Network,* NCC Publications, 1987

Marcham P S, *Data Transmission via PABXs,* NCC Publications, 1984

Nichols E, Jocelyn S, *Selection of Data Communications Equipment,* NCC Publications, 1979

Systems International, Electrical Electronic Press

The Communications User's Yearbook, NCC Publications, 1987

Appendix 3

Summary of CCITT G-series Recommendations

G-series recommendations covering digital networks: transmission systems and multiplexing equipments.

G702	Digital hierarchy bit rates.
G703	Physical/electrical characteristics of hierarchical digital interfaces.
G704	Functional characteristics of interfaces associated with network nodes.
G705	Characteristics required to terminate digital links on digital exchange.
G711	Pulse code modulation (PCM) of voice frequencies.
G712	Performance characteristics of PCM channels between four-wire interfaces at voice frequencies.
G713	Performance characteristics of PCM channels between two-wire interfaces at voice frequencies.
G721	32 Kbit/s adaptive differential pulse code modulation (ADPCM).
G731	Primary PCM multiplex equipment for voice frequencies.
G732	Characteristics of primary PCM multiplex equipment operating at 2048 Kbit/s.
G735	Characteristics of primary PCM multiplex equipment operating at 2048 Kbit/s and offering digital access at 384 Kbit/s and/or synchronous digital access at 64 Kbit/s.

G736	Characteristics of a synchronous digital multiplex equipment operating at 2048 Kbit/s.
G737	Characteristics of an external access equipment operating at 2048 Kbit/s offering digital access at 384 Kbit/s and/or synchronous digital access at 64 Kbit/s.
G741	General considerations on second order multiplex equipments.
G742	Second order digital multiplex equipment operating at 8448 Kbit/s at the third order bit rate of 34,368 Kbit/s and the fourth order bit rate of 139,264 Kbit/s and using positive justification.
G811	Timing requirements at the outputs of reference clocks and network nodes suitable for plesiochronous operation of international digital links.
G821	Error performance of an international digital connection forming part of an integrated services digital network.
G822	Controlled slip rate objectives on an international digital connection.
G823	The control of jitter and wander within digital networks which are based on the 2048 Kbit/s hierarchy.
G921	Digital sections based on the 2048 Kbit/s hierarchy.
G952	Digital line systems based on the 2048 Kbit/s hierarchy on symmetric pair cables.
G954	Digital line systems based on the 2048 Kbit/s hierarchy on coaxial pair cables.
G956	Digital line systems based on the 2048 Kbit/s hierarchy on optical fibre cables.

Appendix 4

Summary of CCITT I-series Recommendations

I-series recommendations covering Integrated Services Digital Network (ISDN).

I110 General structure of the I-series recommendations.

I111 Relationship with other recommendations relevant to ISDNs.

I112 Vocabulary of terms for ISDNs.

I120 Integrated Services Digital Networks (ISDNs).

I130 Attributes for the characterisation of telecommunication services supported by an ISDN and network capabilities of an ISDN.

I210 Principles of telecommunication services supported by an ISDN.

I211 Bearer services supported by an ISDN.

I212 Teleservices supported by an ISDN.

I310 ISDN — Network functional principles.

I320 ISDN — ISDN Protocol reference model.

I330 ISDN numbering and addressing principles.

I331 (E.164) Numbering plan for the ISDN era.

I340 ISDN connection types.

I410 General aspects and principles relating to recommendations on ISDN user–network interfaces.

I411	ISDN user–network interfaces — Reference configurations.
I412	ISDN user–network interfaces — Interface structures and access capabilities.
I420	Basic user–network interface.
I421	Primary rate user–network interface.
I430	Basic user–network interface — Layer 1 specification.
I431	Primary rate user–network interface — Layer 1 specification.
I440	(Q920) ISDN user–network interface data link layer — General aspects.
I441	(Q921) ISDN user–network interface data link layer specification.
I450	(Q930) ISDN user–network interface layer 3 — General aspects.
I451	(Q931) ISDN user–network interface layer 3 specification.
I460	Multiplexing, rate adaptation and support of existing interfaces.
I461	(X.30) Support of X.21 and X.21 bis based Data Terminal Equipments (DTEs) by an Integrated Services Digital Network (ISDN).
I462	(X.31) Support of Packet Node Terminal equipment by an ISDN.
I463	Support of Data Terminal Equipments (DTEs) with V-series type interfaces by an Integrated Services Digital Network (ISDN).
I464	Multiplexing, rate adaptation and support of existing interfaces for restricted 64 Kbit/s transfer capability.

Index

ACE	23, 58-60, 67, 68
Adaptive Differential PCM (ADPCM)	41
A-law	135
analogue communication	13, 14, 27-29, 156-158
baseband	99-102
B-channel	138-140
bit-interleaving	36-39
blocking	66, 71, 72
British Telecommunications plc	17, 18, 22, 28, 59, 125, 135-138, 141, 142, 148, 149
broadband	99, 101, 102
broadband ISDN	138, 139
byte-interleaving	36-39
CCIR	136
CCITT	20, 84, 87, 95, 125, 135, 136, 140-145
CEPT	19, 20
character-interleaving	36-39
Circuit Termination Unit (CTU)	20, 103
codec	76, 77, 127, 128
Continuously Variable Delta Modulation (CVSD)	40
data compression	45

data PBX	57, 62, 63, 68-70
Dataphone Digital Service (DDS)	26
data switching	55-73, 89, 100
Data Terminating Equipment (DTE)	84-89, 97-100, 103, 154-156
D-channel	138-140, 150
Digital Access Signalling System (DASS)	125, 128, 141, 142
digital communications	13-15, 23-30
Digital Private Network Signalling System (DPNSS)	125, 128, 142, 145
Dynamic Time Division Multiplexing	39
Erlang	66, 67
facsimile	147
Fibre Distributed Data Interface (FDDI)	105
flow control	42
Front End Processors (FEPs)	57, 63-65, 70
HDB3	24, 58
HDLC	86
high-speed circuit	12
Huffman encoding	45
IEEE 802 standards	99, 100
integrated communications	14, 28, 39, 48, 49
Integrated Digital Access (IDA)	18, 136, 137, 140-142, 145-150
Integrated Digital Network (IDN)	135, 136
Integrated Services Digital Network (ISDN)	18, 28, 30, 56, 124, 125, 128-131, 133-152, 156, 159
Integrated Services Local Network (ISLN)	105
Integrated Services PABX (ISPABX)	128, 129, 141, 142, 159
I-series recommendations	125, 136, 141-145
kilobit matrix switch	57, 61, 62, 68
kilobit transmission	12, 20, 21, 103, 104
KiloStream	18, 20, 26, 136, 146, 149

INDEX

Local Area Network (LAN)	69, 70, 76, 89, 90, 91, 95, 96, 99-105, 130, 131
low-speed circuit	12
Manchester encoding	101
megabit circuit switch	57-61
megabit transmission	12, 21, 22, 27, 28, 103
MegaStream	18, 21, 22, 27, 136
Mercury Communications Ltd	17-19
Mu-law	134, 135
muldex	24
multiplexing	24, 25, 31-53, 69, 156, 157
network concentrator	57, 65
network management	107-122
Network Terminating Equipment (NTE)	140
Network Termination Unit (NTU)	20, 61, 103, 119, 141
non-blocking	67, 72
non-switched services	20
OSI	86, 95, 125, 142, 145
PABX	57, 62, 65-67, 69, 71-73, 95, 105, 123-131, 141, 156
Packet Assembler/Disassembler (PAD)	46, 85, 87-89, 104
Packet Switched Data Network (PSDN)	83-98, 104
packet switching	83-98
Packet Switching Exchange (PSE)	89-95
Packet SwitchStream (PSS)	18, 92, 137, 148, 149
photo videotex	138, 147
primary multiplex	24, 26
protocol transparency	39, 47, 48
PTT	17-19
Pulse Code Modulation (PCM)	31, 40, 127, 134, 135, 141, 144
RCE	59, 60
Signalling System No 7	145

slow-scan TV	49, 148
SNA	78, 79, 89
Statistical Time Division Multiplexing (STDM)	41-48, 84
Stored Program Control (SPC)	67, 135, 137
switched services	20, 84-87
synchronisation	36, 38
System X and Y	135
T-Carrier system	26, 27
teletex	147-149
telex	34, 89, 147, 149
textfax	148
Time Division Multiplexer	24
Time Division Multiplexing (TDM)	35-41, 47, 48, 127
video conferencing	49, 75-78, 139
videotex	93, 147
viewdata	89, 93
voice	39-41, 46, 104, 105, 126, 127, 131, 156, 158, 159
X.21	21, 86, 140
X.25	85-87